W9-BII-770

Elements of Sampling Theory

Vic Barnett

Professor of Statistics,
University of Bath

Distributed in the United States by
CRANE, RUSSAK & COMPANY, INC.
347 Madison Avenue
New York, New York 10017

ISBN 0 340 17386 6 Boards
ISBN 0 340 17387 4 Unibook

First printed 1974

The English Universities Press Ltd
St Paul's House Warwick Lane London EC4P 4AH

Printed in Great Britain by
Bell & Bain Ltd., Glasgow.

Preface

With the continual expansion of statistical theory, methodology and fields of application, it is inevitable that programmes of instruction in statistics cannot hope to present all aspects of the subject in full detail. A judicious choice of material is necessary. Rather than omit specific areas of study it is preferable, and indeed common, to cover some topics by means of short courses designed to provide a brief introduction to their methods and applications. Such courses present the student with a general survey of the topic and provide a springboard for more detailed individual enquiry, or for more specialised formal study at a later stage.

The collection and processing of data from *finite populations* is an important statistical topic from the practical and utilitarian standpoint, and can be a complex field of study in terms of statistical theory and methodology. Modern society abounds with published and broadcast reports of sample surveys which aim to describe the world we live in. In such surveys, samples are drawn from finite populations and are used to reflect on the population they claim to represent, or indeed are even extended in their claimed import to wider situations. Any student of statistics should have some acquaintance with the principles and methods that are used, and should be aware of the pitfalls of survey sampling. But unless he is specialising in this aspect of statistics it is unreasonable to expect his training to include a comprehensive treatment. Most likely this will be one of the subjects covered by the type of short 'special topics' course referred to above.

The study of *sample survey methods* contains many aspects. It involves setting up appropriate statistical principles, and constructing suitable statistical methods for collecting and analysing data from finite populations. But the implementation of such methods is bound up with sociological, psychological (and other) considerations. To know that a particular method of sampling and estimation is statistically desirable does not necessarily mean that it can be easily applied. Statistical propriety is essential. But such questions as practical access to the population of interest, the social acceptability of an enquiry, personal bias in the response to questionnaires depending on the formulation of the questions, all impose difficult non-statistical considerations. All 'these various aspects have been widely discussed individually in detailed texts. But what appears to be lacking is a concise modern treatment of the subject at an intermediate level. This monograph is designed to form the basis of a short course of instruction, perhaps constituting about 15 lectures. Inevitably it cannot fully scan the field, and it is directed principally to the study of the *statistical* aspects, whilst keeping in mind the problems involved in their application.

The book is based on short lecture courses given, in the Universities of Birmingham, Western Australia and Newcastle upon Tyne, predominantly to senior undergraduate and postgraduate students in statistics, but also on an interdisciplinary basis. It discusses the principles of different methods of probability sampling from finite populations in relation to their relative ease and efficiency for estimating properties of the population of interest. Chapter 1 considers some non-statistical aspects of survey sampling, and sets up the model for probability sampling. Chapter 2 is concerned with simple random sampling as a basis for estimating population means, totals and proportions. In Chapter 3 we discuss ratio and regression estimators which can exploit auxiliary information on additional variables in the population. Chapters 4 and 5 consider situations where further structure exists in the population and simple stratification or clustering methods may be appropriate. Some more complicated probability sampling schemes are described briefly in Chapter 6.

The emphasis is methodological; properties of different sampling schemes and estimators are discussed qualitatively as well as being formally justified. The treatment is at an intermediate level with mathematical proofs being heuristic rather than fully rigorous. A knowledge of elementary probability theory and statistical methods is assumed, such as would be obtained from an introductory course in Statistics. One unifying feature of the book is the empirical study of an actual simple finite population. Throughout the book the application and relative merits of different methods of estimation are demonstrated experimentally by constructing frequency distributions of estimates from this population. This augments and illustrates the theoretical discussion of the various techniques.

It is hoped that, as well as providing a basis for a short course for statistics students, the book may also serve as an introduction to the statistical methods of sampling for those involved in such work at a practical level in various fields of application, including business administration, medicine, psychology and sociology.

I am indebted to the Literary Executor of the late Sir Ronald A. Fisher, F.R.S., and to Dr. Frank Yates, F.R.S., and to Longman Group Ltd., London, for permission to reprint Table 1 from their book *Statistical Tables for Biological, Agricultural and Medical Research*.

It is a pleasure to acknowledge the help of friends and colleagues. I am grateful to David Brook and Betty Gittus for their useful comments on certain sections of the material, to Shiela Boyd for computer calculations, to Ray White for the two cartoons in Chapter 1, and to Shirley Daglish for her careful preparation of the typescript.

Vic Barnett

To Rose

Contents

1 Introduction

One feature of our present society is its preoccupation with numbers. There is a desire to express in quantitative terms all aspects of our lives, from wash-day habits to political ideals. The communications media, including advertisements, television, radio, and the newspapers, maintain a constant flow of such information. No argument is complete without 'the figures to back it up'. Thus we may read (or hear) that:

> *'food prices in the U.K. increased by 7% over the last 4 months'*,
> *'200 million viewers throughout the world watched last night's prize fight on television'*,
> *'2 women out of 10 now use a front-loading washing machine'*,
> *'in 1970 tenants of local authority houses in Greater London paid an average of £3·16 per week in rent'*.

The presentation of such figures is designed to keep us informed of the situation in the world around us; it is often used to 'justify' some proposal or criticism, or at least to place a discussion 'in the proper perspective'. Figures on drinking, or smoking, habits may be presented in support of changes in the traffic laws, or as a partial explanation of different causes of death in different sectors of the community. Results of opinion polls may be advanced as predictions of the outcome of a forthcoming election or as an example of the need to change laws in relation to changing social *mores*.

Undoubtedly the man in the street is better informed than he ever was before, in the sense of being more exposed to quantitative descriptions of the world in which he lives. This is surely a good thing, but it carries contingent responsibilities on both the recipient and exponent of numerical information. On the one hand our 'man in the street' needs to be able to understand and interpret the information presented to him. Some rudimentary knowledge of statistics is obviously desirable (if not common). The ease with which data can be misrepresented or misinterpreted makes one sympathise with the old cry of 'lies, damn lies and statistics'—the prescription *cave emptor* has some justice! But on the other hand, there is a double responsibility on those who present

1

statistical data: to do so fairly and objectively with no intent to deceive, and to provide sufficient detail on the source, scope, and method of collection of the data for proper interpretation or further analysis.

Neither party fully accepts these responsibilities. In spite of an improvement in numeracy, the attitude to statistical data is often one of bemusement or suspicion rather than of understanding or enlightenment. The presentation of data leaves much to be desired. Imprecise or incomplete statements, invalid inferences, graphs with distorted or unspecified scales (or relating to ill-defined concepts with pseudo-scientific names), and pictorial diagrams with psychological impact different from their factual basis all confuse the recipient. Whether such devices are deliberate, or merely arise through lack of statistical expertise, they can certainly serve vested interests to advantage.

All these considerations highlight the need for greater enlightenment in the proper collection, presentation, and interpretation of statistical data. The dangers are illustrated in an entertaining way by Huff (1973). Lay introductions to simple statistical methods are given by Moroney (1951), and Bartholomew and Bassett (1971).

1.1 Finite Populations and Sample Surveys

Such problems of inadequate expression and understanding of statistical data can only be resolved by improved understanding of basic statistical ideas in the population at large. Changes in our educational system are beginning to recognise this, although progress is slow and sporadic. But whilst this need spans all aspects of statistical enquiry, the examples above refer to a rather special type of situation and their study requires a rather special expertise. What characterises them is that they all relate to *finite populations*. The data describe a *limited* and clearly defined (albeit large) set of *individuals* or individual units: the prices of all foods available on the U.K. market, those people throughout the world with access to television presentation, all people who rent houses from the Greater London Council, and so on. The aim is to say something about these finite populations by collecting and analysing information relating (in the main) to only a *part* of that population—what we call a *sample* from the population. This is obtained by *surveying* the population, and the study of how we should reasonably carry out such *sample surveys* is the special topic that we shall consider in this book.

The principles and methods of collecting and analysing data from *finite* populations is a branch of statistics known as *Sample Survey Methods*; their formal basis, *Sampling Theory*. Sample survey analyses are a principal research tool in a vast range of subjects including Agriculture, Education, Industry, Medicine, Psychology, and Sociology. Whether the practitioner is an expert in his own particular subject or a professional statistician, he needs to understand the basic principles and methods that underly the efficient study of finite populations.

The statistician himself may not have encountered in his training any coverage of this topic, because of the pressure of conflicting claims for attention by the vast assortment of different aspects of his subject. In these pages we shall review the elements of sampling theory, at a level which will be appropriate to a student of statistics or to professionals in applied areas with some knowledge of the basic statistical ideas of probability distributions, estimation, and hypothesis testing. If the 'expert' knows what he is doing, we have at least the basis for an improved understanding in the public at large of the import of information obtained from sample surveys.

Let us consider one or two examples from different subject areas, illustrating the range of considerations and difficulties which may arise.

Agriculture

(a) The level of food (for example, meat) prices over recent years has caused some concern. To study the current situation it would obviously not be feasible to determine the prices charged at any time for, say, all cuts of meat by every retail butcher or at every wholesale meat market. But some indication could be obtained by selecting a few cuts of meat and enquiring about prices on a selective basis from different butchers or markets. But how should we effect this choice; what practical difficulties will we encounter; what will our survey tell us of the overall situation?

(b) A county council is required to submit an annual statement of the total wheat yield in its area. With great effort it might attempt a complete ennumeration by contacting every farm. But there is no guarantee that it will receive a full response, or correct information, in each case. Just what is meant by a 'farm' anyway; is this a suitable unit for enquiry? On cost considerations it might again make sense to *sample* on some appropriate basis. But the sample will need to be adequate to meet certain requirements of accuracy and validity—it must reasonably 'represent' the population.

Education

Suppose an enquiry is to be conducted into the attitudes of school-children to the subjects they are studying. Questions of interest might include: 'are they satisfied with their choice of subjects?', 'do they find some subjects more interesting than others?', 'what was the basis of their choice?', and so on. Cost and convenience again suggest a sample survey rather than a total national enquiry. This might be carried out for all schoolchildren in a particular metropolitan region. Hopefully its results would have a wider (possibly national) relevance. Attitudes may be expected to vary with the child's age, intellectual level, chosen subject combinations, and many other factors. An assortment of measures are simultaneously of interest. In the main, these are subjective rather than factual and will need to be elicited by individual enquiries in the form of questionnaires or interviews. The very way in which questions are asked can have a marked effect on the reaction of a child to the enquiry.

Industry

(a) Market research is an important tool in the design of advertising campaigns, in the choice of types of product offered, and in their manner of presentation. Public attitudes to products are commonly sought by means of sample surveys. It will be expected that attitudes will vary from one section of the community to another. Suppose the topic of interest is reaction to the method of packaging of different brands of cigarettes. We might expect that different brands will appeal to different groups: to housewives, businessmen, young people, and so on. It is important that our survey both reflects the views of the different groups and provides a representative coverage of the groups. Some people may resent being approached by interviewers in the street with such an enquiry and refuse to co-operate. *Quota sampling* is a technique commonly employed to cope with such problems of representativity and non-response.

(b) Monitoring the quality of manufactured products is a vast problem for industry. Except in the case of complex or expensive items full inspection cannot be justified. Investigation must inevitably be done on a sample basis, and reliable sampling inspection schemes will be needed.

Sociology

(a) We wish to study the attitudes of 18 year olds to the recent changes in legislation in the U.K. which bestows on them adult status, including the right to vote. A survey is to be conducted of this age group in a city area, relating to those whose permanent address is in the city. Consider some of the difficulties in obtaining a sample for this purpose. Recognising that again we will need to sample the population rather than seek complete information, how are we to identify the members of the population we wish to sample? There is unlikely to be any complete official list of 18 year olds. The Electoral Register will not contain *all* current 18 year olds, in view of its method of notification being tied to some particular date. (In any case, it would be most inefficient to scan such a list in search of a sample of a mere minority of its members.) To stand on a street corner and interview people can quite clearly lead to an imbalanced or unrepresentative sample. Non-response apart, we will undoubtedly miss whole sections of our population—those away from the city street for various reasons, either on a temporary or semi-permanent basis (in hospitals, prisons, on holiday or business elsewhere). Injudicious timing of the enquiry can also lead to imbalance. During the week-day some will be at work, others at school, others unemployed— we may obtain many of the latter! At the weekend, or in the evenings, other sources of imbalance arise. Clearly we would wish our sample to be representative of the different sections of the population. Direct approach to schools, hospitals, prisons, employment agencies, and so on, may be a more reliable and fruitful method of enquiry. But other difficulties now arise, principally concerned with obtaining access to the different sources.

(b) Family income and expenditure surveys are an important sociological guide. Consider some of the practical difficulties that might arise in studying family income in some geographical region. One fundamental matter which generates a deal of discussion is what is meant by a 'family'; is it the same as a 'household'? How do we deal with multiple occupancy in houses, flats, or institutions? What items are included as 'income'? In short, just how do we define our *population* and the *measure* we wish to study? Then again personal resistance to such enquiries, and to the method of implementing a survey, can easily lead to imbalanced or inaccurate results. Door-to-door interviewing during the working day presents an obvious example—results are likely to lead to gross under-assessment. Again, consider the effects of inviting voluntary response to an income survey (or to current political or environmental issues)!

Medicine

Great efforts are constantly being made to improve medical services in relation to organisation, care, and treatment. Information on prevailing experiences and attitudes of patients and administrators is vital. Sample surveys are widely employed. This is an area where responses are likely to be seriously affected by human factors—psychological attitudes to doctors or health workers (veneration or impatience), ignorance, or misunderstanding, can confuse issues and lead to wrong answers. In a recent enquiry about cervical cancer a woman was asked, 'Is your husband circumcised?' She replied, 'Yes doctor, *very*.'

1.2 Practical Problems, and Some Definitions

We have considered these examples at some length, since they highlight various practical difficulties which can arise and point the need for some care in defining basic concepts and principles. We proceed to some definitions and to a brief study of the nature of operational problems which may be encountered in carrying out a sample survey. There is some inconsistency in the way different words are used. The following definitions serve our purpose in this book, but minor differences may sometimes be noticed in other treatments.

Fundamental to our studies is the idea of a *finite population*. *Individuals* (not necessarily human; they might be light bulbs, or farms) in the population have certain measures of interest. For example, our concern may be for the life-times of the bulbs, or the annual wheat yields of the farms. We would like to know certain *characteristics* of the population with respect to some measure—such as the *average* life-time of bulbs in a batch, the *total* wheat yield in Northumberland, or the *proportion* of families with incomes in excess of £2000 during 1972. Occasionally we may be able to derive the exact value of such a characteristic by studying every individual in our population. More often, limited time, money, or access dictates that we should *estimate* the characteristic by studying some smaller group of individuals in the population (a *sample* of its members) and infer the value of the characteristic from the information provided by the sample and by any general knowledge we have about the population. Let us identify the basic concepts a little more fully.

Target Population.   The total finite population about which we require information: for example, all 18 year olds in the U.K.

Study Population.   This is the basic finite set of individuals we intend to study: all 18 year olds whose permanent address is in the metropolitan area of our enquiry; all wheat producers in Northumberland in 1972. It may be (as in the latter case) the same as the target population. Alternatively (as in the former case) it may be a more limited, more accessible population whose properties we hope can be extrapolated to the larger target population.

Population Characteristic. That aspect of the population we wish to measure: the *proportion* of 18 year olds who claim that they will exercise their vote at the next election, the *total* wheat yield in Northumberland in 1972. This expresses some aggregate feature of the population in relation to how it varies from one individual to another. Each individual contributes his component (a number or qualitative description) for some *measure* of interest (voting intention, or wheat yield). Since this can vary from one individual to another we term it the *variable* of interest. The population characteristic will usually be a *total* or *mean* of this variable over the population.

Sampling Units and Sampling Frame. Some of the above examples illustrate how it can be difficult or costly to obtain direct access to the whole target population. For this reason the population *we actually study* (the *study population*) may be more limited in scale. Obviously we hope it has similar features to the target population, and thus provides a balanced reflection of it.

But ambiguities can still remain concerning how we define or obtain access to individuals in the study population. Consider some examples. To investigate wheat yield in Northumberland, we could sample fields or farms in the region (however 'field' and 'farm' are defined), or perhaps sample larger administrative areas. Thus the potential members of the sample, the *sampling units*, can have different forms. They may be fields, farms, or administrative areas. A choice must be made at the outset of the enquiry; it can affect the usefulness of different sampling methods. As a further example, suppose we wish to conduct a survey of family expenditure in some city. Although the 'individuals' in our study population are 'families', some conventional definition of 'family' must be adopted before we can proceed. Even so, there is likely to be no easy means of identifying or sampling such 'family' units. It would be far easier to sample *addresses* and to seek information on families at the chosen addresses. So the addresses become the sampling units, even though the population of addresses is not of essential interest. Then again, in a survey on smoking and bronchitis in elderly people, we might most easily obtain information by approaching a sample of medical practitioners and asking about elderly people who have consulted them over a relevant period. The medical practitioners constitute the sampling units; their elderly patients are *sub-units* (which may be included in full in the survey, or further sampled—see *Cluster Sampling* in Chapter 5). Thus we are not sampling *all* elderly people, but only those who have visited their doctors.

So the source of our sample is the *set of sampling units*. This is called the *sampling frame*. Sometimes the sampling units may be the individual members of the study population. Often this is not so and the sampling frame is a coarser subdivision of the study population, with each sampling unit containing a distinct set of population members.

List.   To use the sampling frame as the raw material from which to draw our sample, we must be able to identify the sampling units. Indeed the sampling frame is chosen with this in mind. At best an actual *list* of all sampling units may exist, such as, for example, the list of city addresses, or the list provided by medical records of all elderly patients visiting their doctors in a given area over a certain period. Such a list makes it particularly easy to choose the sample. But if no tangible list is available for consultation, we must at least set up a conceptual list. For example, in studying meat prices at retail butchers, we may not have a list of butchers to peruse at leisure. Nonetheless our list consists of 'all retail butchers', and we must design our sampling scheme in such a way that it generates data from this list.

Such distinctions are at the root of the practical difficulties of operating sample surveys. Let us summarise some of these difficulties as reflected in the examples of Section 1.1. These include the following (in a roughly hierarchical relationship one to another):

(i)   choice of sampling units where various alternatives exist.

(ii)   discrepancy between the ideal of a target population and the reality of an accessible sampling frame.

(iii)   incomplete or intangible listing of sampling units.

(iv)   implementation of the sample survey. Its organisation and administration involves a complex array of problems of planning, costing, and instruction. Furthermore, we should note the following problems.

(a) If different types of individual exist, our sample should reflect these in a balanced way; there may be different problems in sampling the different types of individual. In the survey of attitudes of 18 year olds, those at school, in hospitals, at work, or unemployed, all present distinct sampling problems with regard to access, cost, and accuracy!

(b) Non-response can contaminate the results of a survey, as can psychological attitudes or inadequate understanding on the part of respondents to interviews or questionnaires.

The resolution of these difficulties must be sought at two levels.

Technical problems of the choice of sampling units, administration of the survey, proper design of questionnaires, or adequate instruction of interviewers, and the like, require experience in a variety of applied disciplines. Detailed knowledge of the special features of the actual area of application of the survey (agriculture, medicine, sociology, etc.) must be combined with advice from the psychologist on questionnaire design or on psychological test procedures, from the sociologist (or other appropriate expert) on the availability of relevant lists and records as a basis for the choice of sampling frame, and perhaps from the computer specialist on the automatic processing of the resulting data. A great deal of organised study, which can take us some way in avoiding the pitfalls, has gone into such matters. But we must ultimately depend on the native good sense of the organisers of a survey in the way in which they exploit the local circumstances and learn from their own experiences. Preliminary pilot studies in advance of the main survey can be a valuable aid.

Such 'technical' problems are largely non-statistical. In contrast, questions of the representativeness of a survey, its validity, the choice of appropriate sampling procedures, methods of estimation of population characteristics (and the properties of these estimators) and legitimate interpretation of the results, all depend vitally on a proper understanding and application of *statistical* ideas. A sound *statistical* basis in the design of a sample survey is vital; 'technical' difficulties of implementation can reduce its effectiveness and must therefore be resolved as far as is possible. On the other hand, a 'technically' perfect survey is not fully exploited if its statistical basis is inadequate; it is virtually *useless* if a total disregard of statistical design considerations makes it impossible to interpret or measure the accuracy of the results. We shall return to this point shortly. *Indeed, it may be viewed as the basic theme of the book*.

Whilst remaining aware of the 'technical' difficulties implicit in conducting a sample survey, we shall henceforth concern ourselves predominantly with the statistical theory and method. Both aspects have been extensively discussed elsewhere in the literature. The following brief list of books offers a cross-section of study of some 'technical' aspects of survey sampling. Many of these books are general texts in their subject area, but include discussion relevant to sample survey issues.

Behavioural Science: Festinger and Katz (1953).
Business: Deming (1960), Slonim (1960).
Interview and Questionnaire design: Festinger and Katz (1953), Hyman (1954), Moser and Kalton (1971), Payne (1951), Atkinson (1971).
Psychological Measurement: Nunnally (1967), Savage (1966).
Sociology: Hyman (1954), Moser and Kalton (1971), Young (1966); also Social and Community Planning (1972).

Some general discussion of problems of identification and implementation appears in several texts more specifically concerned with the statistical aspect, e.g. Hansen, Hurwitz, and Madow (1953). The range of books concerned with *statistical theory and practice* includes the classic text by Yates (1960); Stuart (1962)—a brief non-mathematical treatment; Sampford (1962)—more mathematical with applications to agriculture; Cochran (1963)—an intermediate level mathematical text; Hansen, Hurwitz, and Madow (1953)—detailed and wideranging on 'technical' and statistical matters. Other relevant texts are Johnson and Smith (1969), Kish (1965), Parten (1966), Raj (1968, 1972), Yamane (1967); with their different levels and emphasis represented by their titles (see Bibliography).

1.3 Why Sample?

We have seen how the object of our enquiries is a population consisting of a *finite* number of individuals, on each of whom some measure is observable. We want to characterise the population by some aggregate expression of that measure—perhaps its mean, or total over the population. It is natural to ask 'why not observe every individual in the population and thereby obtain the *exact* answer?' In some cases, where the population is small and easily accessible, this is obviously a sensible policy. If I want to determine how much loose change I have in my pocket, I am hardly inclined to take a sample of the coins and to try to *estimate* the total value! But interest seldom centres on such simple situations. More commonly it really does make sense, for a variety of reasons, to limit our study of the population by merely sampling *some* of its members and using the information gained in this way to *infer* the characteristics of the population as a whole. What are these reasons?

Cost

There will be a limit on the resources, in terms of money, time or effort, that we can apply. This commonly precludes a complete enumeration of the population. There is the need also to counterbalance precision and expense. Cursory inspection of a large number of individuals (possibly even the whole population) may yield, in view of inaccuracies of measurement, far less precise information than that obtained from more careful inspection of some judiciously chosen smaller sample. The use of alternative methods of medical testing provides a good illustration of this effect!

Then again we shall see how, within the limitations imposed by some budget, different methods of sampling can yield (*for the same size of sample*) estimates of dramatically different precision. Finally we shall see how the additional precision which arises from increasing the sample size becomes, typically, less and less valuable in relative cost terms.

Utility

In some instances our sampling units may be destroyed in the process of sampling. Here a complete study of the population is sterile even if we can afford it. There is often no point in knowing all about the population if it no longer exists for the exploitation of our knowledge. Thus a manufacturer of light bulbs, or matches, is not going to test the lifetime of each bulb, or strike each match, to demonstrate the quality of his product. He would have nothing left to sell!

Accessibility

Frequently there is different ease of access to different sampling units. Some may not be observable at all. Again we may be *compelled* to accept only a sample from the population. For example, historical records may be incomplete—temperature or rainfall readings over some period of interest may have been recorded sporadically; contemporary attitudes to some controversial issue may have been incompletely recorded. We can not recreate the circumstances for fuller study.

1.4 How Should We Sample?

This is the obvious next question to ask. Its resolution will require a more formal specification of the finite sampling problem, and of the aims and objectives of a sample survey. This will occupy our attention for the remainder of the book. But we can usefully proceed a little further on a purely intuitive basis. The general aim must be to draw a sample which is an 'honest representation' of the population, and which leads to estimates of population characteristics with as great a 'precision' or 'accuracy' as we can reasonably expect for the cost or effort we have expended.

Various pragmatic or intuitively appealing methods of sampling have been advanced, and are widely applied. Such *ad hoc* methods include the following two.

Accessibility Sampling

The prime stimulus is administrative convenience, and a sample is chosen with sole concern for its ease of access. *We take the most easily obtainable observations*. The pitfalls in terms of lack of 'representativeness' are obvious. Consider the following examples.

(i) To study the sizes of lumps of coal brought to the surface in pit trucks, a few shovelfuls are removed from the top of each truck.

(ii) In an opinion poll on colour prejudice volunteers are sought to answer the questions.

(iii) In an investigation of working habits of married women a door-to-door enquiry is conducted in a middle-class suburban area on a weekday afternoon.

The inevitable shortcomings of such samples as guides to the population as a whole are obvious in these examples. In other situations they may be less obvious—but they may be no less serious!

Judgmental or Purposive Sampling

The attitude here is quite different. Recognising that the population may well contain different types of individual, with differing measures and ease of access, the experimenter exercises *deliberate subjective choice in drawing what he regards as a 'representative' sample.* The results of such a sampling procedure *can* be very good, if the experimenter's intuition or judgment is sound. Judgmental sampling involves the elimination of *anticipated* sources of distortion; but there will always remain the risk of distortion due to personal prejudices, or to lack of knowledge of certain crucial features in the structure of the population. This latter factor is well illustrated by the presence of *unrecognised* correlations between the criteria of choice of the sample and the measure being

" Mrs. SMITH , PERSONALLY SELECTED AS A TYPICAL , ORDINARY HOUSEWIFE "

studied. This arises in the well-known example of sampling ash content in coal by taking some from the edge of *each* of a set of piles of coal in order to obtain a 'representative' sample of the different piles of coal. But the ash content varies with the size of the lumps of coal. So, whilst representing the different piles, the sampling procedure ignores the effect of size. We will tend to pick smaller lumps and the sample will be far from 'representative' in this other crucial respect.

Often judgment and accessibility are combined. For example, in *quota sampling* (see Section 4.5) people may be interviewed in the street in an attempt to obtain a sample judged to be well representative of different ages, sex, occupations, and so on. But this involves an element of accessibility: the 'most promising looking passers-by' are chosen to fill the quotas.

Whilst less prone to dangers than accessibility sampling, the major criticism of judgmental sampling is not that it may lead to unrepresentative samples, but that its results are unconvincing (and too easily dismissed if unpalatable) because there is no yardstick against which to measure 'representativeness' or to assess the propriety of estimators based on such a sampling principle. Such a yardstick is vital!

For this reason we are compelled to introduce an element of 'randomness' into sampling procedures and to draw our samples according to some imposed probability mechanism. A variety of *probability sampling schemes* have been devised, evaluated, compared, and utilised, and we shall study these in some detail as the only sound basis for survey sampling.

To proceed with this we must begin to formalise our finite population model and our sampling objectives.

1.5 The Basic Problem : Probability Sampling

Let us suppose that, in an effort to study some finite target population, we have resolved the matter of the choice of appropriate sampling units and of the sampling frame they comprise. Suppose that the sampling frame represents the accessible finite population, and the sampling units are the individual members of that population. We shall refer merely to the 'population' and its 'members' or 'individuals'. Our interest centres on the values taken by some variable, X, for the different members of the population, and on aggregate measures of this variable over the population. Thus if there are N members, we can represent the population by

$$X_1, X_2, ..., X_N,$$

these being the values taken by X for the different members.

We will be interested in population characteristics defined with reference to X. Those most commonly studied are:

(i) the population *total*, $\quad X_T = \sum_{i=1}^{N} X_i,$

(ii) the population *mean*, $\quad \bar{X} = \frac{1}{N} \sum_{i=1}^{N} X_i = X_T/N,$

and

(iii) the *proportion*, *P*, of members of the population which fall into some category of classification for *X*.

Thus in a social survey of smoking habits in an adult population, *P* may be the proportion of adults who smoke more than 20 cigarettes each day.

The aim of the sample survey will be to *estimate* one or more of these characteristics from the information contained in a sample of $n (\leq N)$ members from the population. Suppose the values of *X* for the sample are

$$x_1, x_2, \ldots, x_n$$

where each x_i is one of the values X_j, of *X*, in the population at large. Not all X_j are necessarily different, so that the same is true of the x_i. Strictly speaking, different x_i *might* arise from the *same* X_j (if we are sampling 'with replacement') but we shall assume (other than in part of Chapter 6) that this is not so. That is, the sample is assumed to have been drawn '*without replacement*'—once an individual has been chosen he cannot be chosen again. (The parallel study of sampling with replacement uses traditional statistical ideas and does not involve many of those special considerations relating to sampling without replacement from a finite population.)

Although *X* may in practice be multivariate (consider the response of an individual to a questionnaire), we shall concentrate on situations where *X is univariate*. Problems of simultaneous estimation of characteristics of the components of multivariate *X*, or of associations between these components, will not be directly considered. But we will later be considering how joint observation of two associated variables may lead to more efficient estimation of the characteristics of one of them, and how to estimate ratios of characteristics of the components of such bivariate *X*. (See Chapter 3.)

Terminology. The ratio of sample size to population size

$$f = n/N$$

will be called the *sampling fraction*.

To estimate X_T, $\bar{X}$, or *P* we will calculate some summary measure of the sample. Thus to estimate $\bar{X}$ it might seem appealing to use the *sample mean*

$$\bar{x} = \frac{1}{n} \sum_{i=1}^{n} x_i.$$

But how are we to assess the propriety of such an *estimator*? One possibility is to enquire how values of $\bar{x}$ may vary in relation to $\bar{X}$ from one occasion to another when we employ the current sampling procedure in the same problem. With accessibility or judgmental sampling we cannot answer such a question; the lack of any objective sampling principle rules it out.

Consequently we are led to introduce some probability mechanism as a means of drawing samples, and must consider the idea of:

Probability Sampling

In general terms, we firstly specify the size, n, of sample to be drawn. We then consider (conceptually at least) *all* possible samples of size n; $S_1, S_2, \ldots$: that is, each S_i is a *distinct* sample of size n *drawn from the whole population.*

A probability sampling scheme then assigns a probability π_i to each S_i and a particular sample, S, is chosen in accord with this probability structure. A vast assortment of different probability sampling schemes are possible, corresponding to different probability distributions $\pi = \{\pi_1, \pi_2, \ldots\}$. These range from simply defined and implemented schemes to highly sophisticated ones. We shall be considering an assortment of the more straightforward schemes that are commonly used, and comparing them in terms of their cost and their efficiency for the estimation of $\bar{X}$, X_T, and so on.

With such a probability-based sampling principle we are now able to discuss the 'representativeness' of the sample (in terms of its method of generation) and the 'accuracy' of estimators, employing the usual statistical concepts.

Suppose that θ is some population characteristic (it may be X_T) and we choose to estimate it by some function, $\tilde{\theta}(S)$, of the sample. $\tilde{\theta}$ is called a *statistic* or *estimator*. We can discuss the properties of the sampling scheme and of the estimator in terms of the *sampling distribution* of $\tilde{\theta}$ induced by the probability distribution π. Different values of $\tilde{\theta}$ will be encountered on different occasions, with probabilities determined by π.

Unbiasedness.   One possible criterion on which the sampling scheme may be judged 'representative' is the $\tilde{\theta}$ should be unbiased. That is

$$E_{\underset{\sim}{\pi}}[\tilde{\theta}(S)] = \theta,$$

where E is the expectation operator.

We shall give prime attention to unbiased estimators. Only occasionally are we prepared in sample survey work to 'trade bias for precision'.

Precision. Often the estimator $\tilde{\theta}$ has, at least in large samples, an approximately normal distribution. It is reasonable, therefore, to assess the 'accuracy' or 'precision' of an unbiased estimator by considering its *variance*,

$$\text{Var}\ [\tilde{\theta}(S)] = E_{\underset{\sim}{\pi}}\{[\tilde{\theta}(S)-\theta]^2\}$$

—the smaller this variance, the more 'precise' the estimator. If, for a given size of sample, one unbiased estimator has lower variance than another, we say it is *more efficient*. In this way we can compare estimators or different probability sampling schemes.

If $\tilde{\theta}$ is biased, Var $[\tilde{\theta}(S)]$ needs to be replaced by the *expected mean square error*,

$$\text{E.M.S.E.}\ [\tilde{\theta}(S)] = E_{\underset{\sim}{\pi}}\{[\tilde{\theta}(S)-\theta]^2\}$$

—the smaller this quantity, the better (or more 'precise') the estimator. Occasionally, a biased estimator with small E.M.S.E. may be preferred to an unbiased one with larger variance, but we shall concentrate attention on unbiased estimators in the main.

The broad aim of sampling theory is to devise sampling schemes which are economical and easy to operate, which yield unbiased estimators, and which minimise the effects of sampling variations. This latter factor is reflected by Var $[\tilde{\theta}(S)]$ for an unbiased estimator. In general Var $[\tilde{\theta}(S)]$ decreases with increase in sample size, but the cost increases. We must effect a balance. We shall be comparing sampling schemes to determine which of them yields an unbiased estimator with smallest variance *for a given cost*, or *for a given sample size*. We must also tackle the inverse problem of choosing the sample size to yield prescribed precision (in terms of $\{\text{Var}\ [\tilde{\theta}(S)]\}^{-1}$). The ease of operation and administration of a sample survey is an important aspect of cost reduction. We shall sometimes find that sheer administrative convenience can override other factors in promoting a particular method of sampling. (See *cluster sampling*, *systematic sampling*, and *quota sampling*.)

1.6 "Class of '68" : A Finite Population

To illustrate the theoretical results obtained at later stages, we shall simultaneously construct empirical sampling distributions from an actual finite population. The population consists of all 25 members of a university class of statistics students who, in 1968, were taking a short course in Sampling Theory. The variables recorded were the weight, height, and sex of each member of the class. Heights and weights are denoted X and Y, and recorded in inches (in excess of 60 in) and pounds (in excess of 100 lb), respectively. Figure 1 presents the details of this population, including their seating positions in the first lecture, with rows labelled I–V, and columns labelled A–E.

Whilst the population is a genuine one, it is not suggested that the procedures subsequently applied to this population, or the assumptions made at different stages, are realistic in practical terms. We would not apply such ideas to such a small population in any case. Its study is purely illustrative, serving to endorse the formal results obtained in later sections.

Example

Suppose we adopt one or other of the two intuitive sampling methods described above to draw a sample of size 5 from the *Class Example* of Figure 1, for the purpose of estimating the mean height of the class. No prescription can be given for drawing an accessible or judgmental sample. The type of sample chosen will vary with the whim of the sampler. However, to some, a fairly obvious *accessible* sample is constituted by the first row. The sample mean is

$$\bar{x}_I = 17/5 = 3 \cdot 4,$$

a very low value in comparison with the population mean, $\bar{X} = 8 \cdot 44$. But this is hardly surpising; all members of the first row are female—compared with the predominance of males in the population as a whole.

In drawing a *judgmental* sample we consciously attempt to avoid such 'unrepresentative' samples. We might choose one member from each row, say, by taking the diagonal from

Population: Members of a class of students (1968–69).

Variables Sex   (M or F)
             Height (X_i, inches in excess of 60 in)
             Weight (Y_i, lb in excess of 100 lb).

M or F
X_i
Y_i

Variable of principal interest: *Height.*

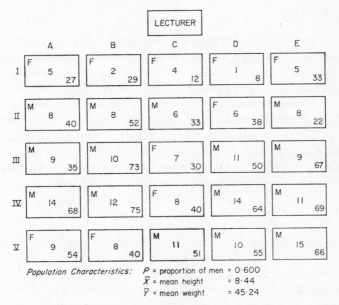

Population Characteristics: P = proportion of men = 0·600
                             $\bar{X}$ = mean height   = 8·44
                             $\bar{Y}$ = mean weight   = 45·24

Figure 1 Class Example.

IA to VE. This has 60 % male members as required, and the sample mean is

$$\bar{x}_J = 9\cdot8,$$

which is somewhat closer to 8·44. But we have no way of assessing the sampling procedure in terms of values of $\bar{x}_J$ which might arise. Presumably, if the chosen sample has the maximum appeal to the sampler, he will always choose this sample. Thus he *always* obtains 9·8 as his estimate. So the sampling procedure will lead to a good estimate if it leads to a good estimate, and vice versa! We can say no more regarding precision.

Exercises for Chapter 1

I

For the illustrative examples, *Agriculture* (a) and *Education* in Section 1.1 discuss the difficulties underlying the choice of an appropriate sampling frame, and the organisational problems likely to be encountered in collecting data from your preferred frames.

II

In a national enquiry into Health Service needs, it is necessary to study, through an appropriate sample survey, the pattern of general medical practice. We want to examine the 'referral' pattern for general practitioners over a particular year: how many patients they send to hospital for inpatient or outpatient treatment; how many to consultants for specialist comment, or to psychiatrists; how many need the services of health visitors; and so on. Two alternative sampling schemes are available: complete enumeration for a small sample of practitioners, or a sample of individual patients throughout the country. Discuss the possible advantages and disadvantages of the two schemes.

III

We wish to examine the views of a class of undergraduate university students on a lecture course they have taken on Sample Survey Methods. The aim is to use this feedback of information to redesign the course (if necessary) in as far as the method of presentation of the material effects its impact and understanding. Construct a short questionnaire for circulation to the members of the class for this purpose.

IV

(*For class co-operation*). Choose an *accessible* and a *judgmental* sample of size 5 from the *Class Example* for the purpose of estimating mean weight. Contribute your estimates to the results for the whole class and construct histograms in the two cases. Discuss what appear to be the factors which have influenced the choice of samples by the two methods.

V

Suppose it is necessary to draw a sample:

(i) of telephone subscribers whose telephone numbers refer to a particular exchange, from a directory which covers many exchanges in addition to the one of interest,

or

(ii) of a particular set of authors, from a composite list of all books published by those authors.

Discuss any difficulties which might arise in drawing reasonable samples in each case, and suggest methods of overcoming such difficulties.

2 Simple Random Sampling

The most basic form of probability sampling is *simple random* (s.r.) *sampling*. It is widely used in its own right and is easy to operate from the statistical viewpoint. It also serves as the basis for more complicated sampling schemes, such as *stratified simple random sampling*, and *cluster sampling*. The properties of estimators obtained from simple random samples may be readily demonstrated, and this will be the object of the present chapter.

2.1 The Simple Random Sampling Procedure

This operates in the following way. If the population is of size N, and we require a *simple random sample* of size n, this sample is chosen *at random* from the $\binom{N}{n}$ distinct possible samples, in each of which no population member is included more than once. That is to say, each of the $\binom{N}{n}$ samples has the same probability $\binom{N}{n}^{-1}$ of being chosen. Such a sample can be obtained sequentially: by drawing members from the population one at a time *without replacement**, so that at each stage every remaining member of the population has the same probability of being chosen.

We see that this produces a *simple random sample* as follows. Suppose this sequential method of choice yields n (distinct) population members whose X values are

$$x_1, x_2, \ldots, x_n$$

where x_i refers to the ith chosen member ($i = 1, 2, \ldots, n$).

The probability of obtaining this *ordered* sequence is

$$\frac{1}{N} \cdot \frac{1}{N-1} \cdots \frac{1}{N-n+1} = \frac{(N-n)!}{N!}$$

But any reordering of $x_1, x_2, \ldots, x_n$ corresponds to the same choice of n distinct population members; there are $n!$ possible reorderings. Thus

---

* Throughout the book, the term 'simple random sampling' will be used to describe sampling at random *without replacement*; the lack of replacement is implicit in the definition.

the probability of obtaining any particular set of n distinct population members (irrespective of order) is just

$$\frac{n!\,(N-n)!}{N!} = \binom{N}{n}^{-1} .$$

There are $\binom{N}{n}$ such sets (or *samples*) which can arise, and these samples are therefore generated with equal probabilities: that is, *by simple random sampling.*

The choice of individual observations in the sample is achieved at each stage by an appropriate random mechanism applied to the remaining members of the population; for example, using a table of random digits such as that given as Table 1 in the Appendix.

Example 2.1

Choose a simple random sample of 5 heights from the Class Example given in Figure 1. Numbering the population members 0 to 24 along the rows, starting from the top left hand corner, we find that the first five distinct pairs of numbers less than 25 in Table 1 (reading along the rows) are 23, 24, 19, 09 and 06 yielding a simple random sample 10, 15, 11, 8, 8, of heights.

In such a simple situation this is a reasonable task. But notice how for larger populations and samples the effort of choosing the sample from a table of random digits can be quite excessive. If N is not 10, 100, 1000, and so on, we must either disregard a possibly large portion of the table, or allow multiple reference from the table to individual population members. We must also keep an account of those members which have been chosen and ignore them on subsequent random appearance. These factors make it desirable to give careful thought to the actual mechanics of using the table of random numbers to choose the sample, in order to keep the effort as low as possible.

We shall consider the use of simple random samples for estimating the three population characteristics: the population mean $\bar{X}$, the population total X_T, and the proportion, P, of X values in the population which satisfy some condition of interest. We shall need to discuss how any estimators behave in an aggregate sense: that is, in terms of their sampling distributions. By analogy with the study of traditional estimators of parameters in infinite population models, where the variance of the parent population is often a crucial measure, we need to define the *variance* of a finite population.

Variance. The variance of the finite population X_1, X_2, ..., X_N is

$$S^2 = \frac{1}{N-1} \sum_{i=1}^{N} (X_i - \bar{X})^2.$$

Notice that this is a *deterministic* measure, not a probabilistic average as it is in the case of random variable theory for probability models. The divisor (here $N-1$) is arbitrary. In other treatments of sampling theory sometimes $N-1$ is used, elsewhere N. All we need is some convenient measure of the variability of the X values in the population. S^2 is particularly convenient in that it leads to simpler algebraic expressions in the later discussion, and produces results more closely resembling corresponding ones in the infinite population context.

Probability averaging only arises in relation to some prescribed probability sampling scheme. Thus for simple random sampling we have the concept of the *expected value* of x_i, the ith observation in the sample. This is

$$E[x_i] = \sum_{j=1}^{N} X_j Pr(x_i = X_j) = \frac{1}{N} \sum_{j=1}^{N} X_j = \bar{X}.$$

The result that $Pr(x_i = X_j) = 1/N$ holds because the number of samples with $x_i = X_j$ is $(N-1)!/(N-n)!$, and each has probability $(N-n)!/N!$

We easily see that

$$E(x_i^2) = \frac{1}{N} \sum_{j=1}^{N} X_j^2,$$

and

$$E(x_i x_j) = \frac{2}{N(N-1)} \sum_{r<s} X_r X_s \ (i \neq j).$$

Hence the *variance* of x_i, and *covariance* of x_i and x_j, are

$$\begin{aligned}
\text{Var}(x_i) &= E\{(x_i - \bar{X})^2\} \\
&= E(x_i^2) - \bar{X}^2 \\
&= (N-1)S^2/N
\end{aligned} \tag{2.1}$$

and

$$\text{Cov}(x_i\, x_j) = E\{(x_i - \bar{X})(x_j - \bar{X})\}$$

$$= E(x_i\, x_j) - \bar{X}^2$$

$$= \frac{1}{N(N-1)}\{(\sum_{j=1}^{N} X_j)^2 - \sum_{j=1}^{N} X_j^2 - N(N-1)\bar{X}^2\}$$

$$= -S^2/N. \tag{2.2}$$

Thus, as we might anticipate, there is a small negative correlation between the potential sample observations.

The result (2.1) seems to suggest that the divisor N, not $(N-1)$, would be better in the definition of the finite population variance—for the sake of tidiness. But later results outweigh this, and show that the adopted form, with divisor $(N-1)$, is indeed more convenient.

We can now proceed to study the estimation of the population mean $\bar{X}$.

2.2 Estimating the Mean, $\bar{X}$

An estimator of $\bar{X}$, based on a s.r. sample of size n, with immediate intuitive appeal is the *sample mean*,

$$\bar{x} = \frac{1}{n}\sum_{i=1}^{n} x_i.$$

Let us consider some properties of $\bar{x}$ as an estimator of $\bar{X}$.

We see that

$$\boxed{E(\bar{x}) = \bar{X}, \text{ so that } \bar{x} \text{ is } unbiased,}$$

for, $E(\bar{x}) = \dfrac{1}{n} E(x_1 + x_2 \ldots + x_n) = n\bar{X}/n = \bar{X}.$

Also,

$$\boxed{\text{Var}(\bar{x}) = (1-f)\, S^2/n,} \tag{2.3}$$

B

for,

$$\text{Var}(\bar{x}) = \frac{1}{n^2} \sum_{i=1}^{n} \text{Var}(x_i) + \frac{2}{n^2} \sum_{r<s} \text{Cov}(x_r x_s)$$

$$= \frac{1}{n^2} \{n(N-1)S^2/N - n(n-1)S^2/N\}$$

$$= \left(\frac{N-n}{N}\right)S^2/n = (1-f)S^2/n.$$

We see in (*2.3*) the effect of the population being finite. The sampling variance of $\bar{x}$ is reduced by a factor n/N, the *sampling fraction*, compared with the analogous result for an infinite population. This effect is known as the *finite population correction* (f.p.c.). If the sampling fraction is small, the f.p.c. has little importance, and we can often ignore it. As a rule of thumb we might ignore the f.p.c. if f is less than about 0·05. The consequence is to slightly over-state the variance of the estimator, $\bar{x}$, but such implied conservatism in any assessment of the accuracy of estimation of $\bar{X}$ is usually unimportant.

Terminology.   The *standard deviation* of $\bar{x}$, $[\text{Var}(\bar{x})]^{\frac{1}{2}}$, will be called its *standard error*.

So we can say of $\bar{x}$ that it is unbiased as an estimator of $\bar{X}$, and (*2.3*) enables us to compare it on efficiency grounds with other estimators of $\bar{X}$ based on s.r. samples, or samples obtained from other sampling schemes. Also $\bar{x}$ is *consistent* in the finite population sense: as $n \to N$, so $\bar{x}$ becomes $\bar{X}$. Under appropriate circumstances we can also invoke the approximate *normal* distributional form for $\bar{x}$ to derive confidence intervals for $\bar{X}$, or to choose a sample size n to meet prescribed accuracy requirements. But before considering these matters let us pose a more basic question.

Within the simple random sampling scheme, how well does $\bar{x}$ compare with other possible estimators of $\bar{X}$? One property is easily demonstrated.

The sample mean, $\bar{x}$, is the best linear unbiased estimator of $\bar{X}$ based on a simple random sample of size n.

The expression 'best' is used here to mean 'having smallest variance'. The result does not imply any global optimality among all estimators, $\theta(\mathbf{x})$, of $\bar{X}$, but it is useful to know that in the class of easily calculated *linear unbiased* estimators, the simple form $\bar{x}$ is best.

Let us confirm that this is so. Consider any linear unbiased estimator,

$$t = \sum_{i=1}^{n} a_i x_i,$$

with

$$\sum_{i=1}^{n} a_i = 1$$

to ensure unbiasedness.

$$\text{Var}(t) = \left(\frac{N-1}{N}\right) S^2 \sum_{i=1}^{n} a_i{}^2 - \frac{2S^2}{N} \sum_{r<s} a_r a_s = S^2 \left(\sum_{i=1}^{n} a_i{}^2 - 1/N\right)$$

Thus we need

$$\sum_{i=1}^{n-1} a_i{}^2 + \left(1 - \sum_{i=1}^{n-1} a_i\right)^2$$

to be a minimum, which will be so if

$$a_i = 1 - \sum_{i=1}^{n-1} a_i = a_n.$$

In other words a_i must be constant $(i = 1, 2, \ldots n)$, and the unbiasedness condition shows that, for minimum variance,

$$a_i = 1/n,$$

yielding the sample mean, $\bar{x}$, as the *best linear unbiased estimator*.

2.3 Estimating the Variance, S^2

The expression (2.3) for $\text{Var}(\bar{x})$ is used in three ways:

    (i) to assess the precision of the estimator $\bar{x}$,
   (ii) to compare $\bar{x}$ with other estimators of $\bar{X}$,
  (iii) to determine the size of sample needed to yield a desired precision.

Typically, however, we will not know the value of S^2, so that to make use of the sampling variance (2.3) we must estimate S^2 from sample data. Using the simple random sample $x_1, x_2, \ldots, x_n$ we might try (cf. infinite populations) using

$$s^2 = \frac{1}{n-1} \sum_{i=1}^{n} (x_i - \bar{x})^2.$$

It turns out that

$$E(s^2) = S^2,$$

for,

$$E(s^2) = \frac{n}{n-1} \left\{ \sum_{j=1}^{N} X_j^2/N - E(\bar{x}^2) \right\}$$

$$= \frac{n}{n-1} \left\{ \sum_{j=1}^{N} X_j^2/N - (1-f)S^2/N - \bar{X}^2 \right\}$$

$$= \frac{n}{n-1} \left\{ \frac{N-1}{N} - \frac{1-f}{n} \right\} S^2$$

$$= S^2.$$

So that s^2 is unbiased for S^2.

In relation to problems (i) and (ii), above, we can substitute for the unknown population variance in (2.3) the unbiased sample estimator, s^2, and we have an *unbiased estimator of* $\mathrm{Var}(\bar{x})$ *as*

$$s^2(\bar{x}) = (1-f)s^2/n.$$

Occasions may also arise where estimation of S^2 is of interest in its own right; again s^2 serves for this purpose. But the problem (iii), where we wish to determine the size of sample needed to achieve a desired precision, is less straightforward if S^2 is not known. The sample estimator s^2 is now of little value since we do not have a sample from which to extract it! We need to determine the required sample size *prior to sampling*, and we shall consider later in Section 2.5 how to face up to the difficulty of an unknown population variance.

2.4 Confidence Intervals for $\bar{X}$

Example 2.2

Suppose that for the *Class Example* of Section 1.6 we decide to estimate the mean height $\bar{X}$ from the sample mean of a simple random sample of size 5. Thus if our sample is 10, 15, 11, 8, 8 the estimate of $\bar{X}$ is

$$\bar{x} = 10 \cdot 4.$$

From (2.3) we see that the sampling variance of the sample mean is

$$(1 - 5/25)S^2/5$$

where the population variance for the X-values is in fact
12·42. Thus the standard error of $\bar{x}$ is $0\cdot4S = 1\cdot410$. So the
estimate 10·4 turns out to be about 1·39 standard errors in
excess of $\bar{X}$, which we know to be 8·44.

But what of the actual distribution of $\bar{x}$? We can obtain
some indication of its form from taking repeated s.r. samples
of size 5. Five hundred such samples generated on a computer
yielded values of $\bar{x}$ represented by the histogram* shown in
Figure 2. The sample variance of the 500 sample means is
1·94, which compares well with the expected value 1·99. (The
average of the 500 sample means is 8·46.)

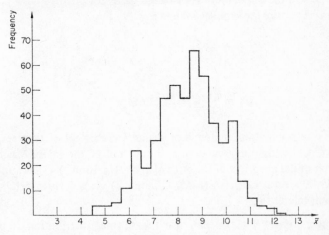

Figure 2   Histogram of 500 values of $\bar{x}$ from s.r. samples of size 5 in the
Class Example.

We see in Figure 2 that, even for such a small sample size and non-
negligible sampling fraction, the sampling distribution of $\bar{x}$ shows no
substantial lack of symmetry. But this effect is assisted by the fact that
the finite population of X-values itself shows little 'skewness'. This will
not always be true of course. We will frequently encounter skew
populations in practice, often positively skew in the sense of exhibiting
a long tail of large X-values. Consider, for example, the numbers of
children in different families, or family annual incomes. Even more
extreme situations will be encountered, with a maximum frequency at
the lowest X-values so that we have an almost i-shaped distribution of

*For this histogram and for the others throughout the book, the data have been
grouped in intervals of width 0·4. The ordinate label 'frequency' relates throughout
to the numbers of observations in such equal-sized intervals.

population values. Consider, for example, the numbers of claims on an insurance policy arising from a population of insured people, or the number of times a medical practitioner sees each of the patients on his panel over a given year.

But, supported by a finite population analogue of the *Central Limit Theorem*, it can often be assumed that the s.r. sample mean has a *normal sampling distribution*. That is,

$$\bar{x} \sim \underset{\sim}{N}(\bar{X}, (1-f)S^2/n). \qquad (2.4)$$

This assumption is often very reasonable even in the presence of skewness in the population. As a rough guide for positive skew populations, we require the sample size n to satisfy

$$n > 25G_1^2$$

where,

$$G_1 = \sum_{i=1}^{N} (X_i - \bar{X})^3/NS^3$$

(the finite population analogue of Fisher's coefficient of skewness). In addition, the sampling fraction, f, should not be too large. Much discussion of the propriety of the normal approximation (2.4) has appeared in the literature. Cochran (1963, Chapter 2) gives further details and relevant references.

Where appropriate we can use the normal distribution to make further inferences about $\bar{X}$. We might wish to construct a confidence interval for $\bar{X}$. An approximate $100(1-\alpha)\%$ *symmetric two-sided confidence interval* for $\bar{X}$ can be written

$$\bar{x} - z_\alpha S\sqrt{[(1-f)/n]} < \bar{X} < \bar{x} + z_\alpha S\sqrt{[(1-f)/n]}, \qquad (2.5)$$

where z_α is the double-tailed α-point of $N(0, 1)$. That is, if

$$Z \sim \underset{\sim}{N}(0, 1); \quad Pr\{|Z| > z_\alpha\} = \alpha.$$

But S will not be known in practice. Replacing S by the sample estimate, s, will be reasonable, provided n is sufficiently large. By analogy with the infinite population case, a better allowance can be made for not knowing S by using *Student's t-distribution* rather than the normal distribution, when n is small (less than about 40).

We then have an approximate $100(1-\alpha)\%$ *symmetric two-sided confidence interval* for $\bar{X}$ in the form

$$\bar{x}-t_{n-1}(\alpha)s\sqrt{[(1-f)/n]} < \bar{X} < \bar{x}+t_{n-1}(\alpha)s\sqrt{[(1-f)/n]} \quad (2.6)$$

where $t_{n-1}(\alpha)$ is the double-tailed α-point of t with $(n-1)$ degrees of freedom. Values of z_α, and of $t_{n-1}(\alpha)$ for a range of values on n are given in Table 2 of the Appendix, for $\alpha = 0\cdot1$, $0\cdot05$, $0\cdot02$, $0\cdot01$, $0\cdot002$ and $0\cdot001$. If n is more than about 40, the normal distribution percentage points will usually be reasonable.

Since sample surveys commonly relate to very large populations (say $N = 10000$ or more) with substantial sample sizes (say $n = 100$ or more), we will frequently be able to safely adopt the form (2.5), replacing S by s, without regard to the fine details of justifying the *normal* distributional form for $\bar{x}$ or serious concern for the sampling fluctuations of s as an estimate of S. However, one word of warning is needed on this latter issue. The sampling variance of s^2 is highly sensitive to the value of the *fourth* central moment of the X values in the finite population. Typically, the larger this fourth moment the larger is $\mathrm{Var}(s^2)$. This is particularly serious when we attempt to compare the precision of alternative estimators and need to estimate S^2, or, even more so, when we need to choose a sample size to yield a required precision (as described in the next section).

Example 2.3

> In a particular sector of industry a survey is conducted in an attempt to investigate the extent of absenteeism not connected with illness or official holidays. A random sample of 1000 men out of a total workforce of 36000 are asked how many days they have taken off work, in the previous six months, as 'casual holidays'. The results were as follows.

'Days off'	0	1	2	3	4	5	6	7	8	9
No. of men	451	162	187	112	49	21	5	11	2	0

> To estimate the average number, $\bar{X}$, of days 'casual holiday' taken by workmen in the industry we can use the sample mean
>
> $$\bar{x} = 1\cdot296.$$

The sample variance is

$$s^2 = 2\cdot397.$$

Using the normal approximation to the distribution of the sample mean we can obtain an approximate 95% symmetric two-sided confidence interval for $\bar{X}$ as

$$1 \cdot 201 < \bar{X} < 1 \cdot 391$$

[or, $1 \cdot 200 < \bar{X} < 1 \cdot 392$ (ignoring the f.p.c.)].

Note some characteristic features of this problem. The distribution of values in the population is obviously highly skew. This will affect the propriety of the normal approximation in general, although the sample size of 1000 here is an adequate safeguard. Then there is the inevitable problem of assessing the accuracy of the information given on such a controversial issue. (Promises of confidentiality may not necessarily allay all concern!) We note that there is no 'non-response' category. If the survey involved compulsory response, fears for its accuracy become more acute. Suppose there had been a 'non-response' category. How would we then process the data to estimate $\bar{X}$?

2.5 Choice of Sample Size, *n*

Clearly an increase in sample size will lead to an increase in the precision of $\bar{x}$ as an estimator of $\bar{X}$—but the sampling costs will also typically increase and there is likely to be some limit on what we can afford. Too large a sample will imply a waste of resources; too small a sample is likely to produce an estimator of inadequate precision. Ideally we should state the precision we require, or the maximum cost which can be expended, and choose the sample size accordingly.

Such an aim involves a complex array of considerations. What is the cost structure for sampling in a given situation; how do we assess the precision we require of our estimators; how do we balance needs in relation to *different* population characteristics which may be of interest; how do we deal with lack of knowledge of unknown parameters (e.g. the population variance) which affect the precision of estimators?

We will consider only one simple situation. We assume that the object is to estimate a single characteristic, the population mean $\bar{X}$, by using a s.r. sample mean $\bar{x}$, restricting to an acceptable level the probability that the absolute difference between $\bar{X}$ and $\bar{x}$ is greater than some specified value. No direct consideration of costs arises, although, if it happens that sampling costs are directly proportional to sample size, it turns out that we achieve our aim for minimum cost.

Suppose we seek the minimum value of n that ensures that

$$Pr\{|\bar{X}-\bar{x}| > d\} \leqq \alpha \qquad (2.7)$$

for some prescribed d and (small) α. The sampler needs to specify the tolerance d, and the risk α of not obtaining such tolerance.

We can rewrite (2.7) as

$$Pr\left\{\frac{|\bar{X}-\bar{x}|}{S\sqrt{[(1-f)/n]}} > \frac{d}{S\sqrt{[(1-f)/n]}}\right\} \leqq \alpha, \qquad (2.8)$$

so that, using the normal approximation (2.4), we require

$$\frac{d}{S\sqrt{[(1-f)/n]}} \geqq z_\alpha$$

or,

$$n \geqq N\left\{1+N\left(\frac{d}{z_\alpha S}\right)^2\right\}^{-1}. \qquad (2.9)$$

Equivalently, (2.8) declares that

$$\text{Var}(\bar{x}) \leqq (d/z_\alpha)^2 = V \text{ (say)}.$$

The inequality (2.9) can be written

$$n \geqq S^2/V\left[1+\frac{1}{N}S^2/V\right]^{-1},$$

so that as a *first approximation* to the required sample size, we could take

$$n_0 = S^2/V. \qquad (2.10)$$

This is an *overassessment*, but it will be reasonable unless the provisional sampling fraction, n_0/N, is substantial. If this is so, we would need to reduce our assessment of the *required sample size* to

$$n_0(1+n_0/N)^{-1}.$$

This presupposes, however, that S^2 is known. Such a possibility is remote in practice and we must face the more difficult task of estimating the minimum sample size required to satisfy (2.7) when S^2 is unknown. There are basically 4 ways in which we might try to do this.

(i) From Pilot Studies

Occasionally a pilot study may be conducted prior to a major sample survey. This can serve a variety of purposes including the study of different sampling frames and the examination of any implicit technical

B*

difficulties which may be encountered in the sampling procedure. If such a pilot study itself takes the form of a simple random sample, its results may give some indication of the value of S^2 for use in the choice of the sample size of the main survey; if the pilot sample is not obtained by a probability sampling procedure we must be circumspect in such an application of the results. For convenience, a pilot study is often restricted to some limited part of the population. If so, the estimate of S^2 which it yields can be quite biased.

(ii) From Previous Surveys

It is not uncommon to find that other surveys have been conducted elsewhere which have studied *similar* characteristics in *similar* populations. This is particularly common in educational, medical, or sociological investigations—it may just be that a different age-group of pupils, or results for a different year, or effects in a different city or social structure, are being considered. Often the measure of variability from earlier surveys can be applied as an indication of S^2 for the present population in order to facilitate the choice of the required sample size to meet any prescription of precision in the current work. But again the natural precautions must be taken in extrapolating from one situation to another.

(iii) From a Preliminary Sample

This is the most reliable approach, but it may not be feasible on administrative or cost considerations. It operates as follows. A preliminary s.r. sample of size n_1 is taken and used to estimate S^2 through the sample variance s_1^2. We aim to ensure that n_1 is inadequate to achieve the required precision, and augment the sample with a further s.r. sample of size $(n-n_1)$, where $(n-n_1)$ is chosen by using s_1^2 as the necessary preliminary estimate of S^2.

A detailed study of this procedure shows that, under reasonable conditions, the total sample size, *ignoring the f.p.c.*, needs to be

$$(1+2/n_1)s_1^2/V$$

an essential increase by the factor $(1+2/n_1)$ over what would be needed if S^2 were known.

This approach, if feasible, is undoubtedly the most objective and reliable. The sampling procedure is an example of what is called *double* (or *two-phase*) *sampling*.

(iv) From Practical Considerations of the Structure of the Population

Occasionally we will have some knowledge of the structure of the population which throws light of the value of S^2. Suppose we are considering
  (a) the numbers of misprints in books (of roughly the same size) issued by a particular publisher over a certain period of time, or
  (b) the number of faults that occur in transistor radios of a particular type in the first year of their use.

In both cases there is reason to believe that the X values might vary roughly in the manner of a Poisson distribution, so that it is plausible to assume that S^2 is of the same order of magnitude as $\bar{X}$. Any information we have about the possible value of $\bar{X}$ (for example, from other similar studies) can then be used to approximate S^2 and assist in the choice of the required sample size.

Then again, if we are interested in estimating a proportion P we shall see that the sampling variance of the s.r. sample estimator is simply related to P. Reasonable bounds can be placed on this variance to obtain some idea of the required sample size. We shall return to this point later (Section 2.9).

2.6 Estimating the Population Total, X_T

There are many situations in which we are interested in estimating the population total

$$X_T = N\bar{X}$$

rather than the population mean $\bar{X}$. For example, in a survey of annual yields of wheat for farms in Northumberland, the concern may be to estimate the county's total annual wheat yield. In view of the simple relationship between X_T and $\bar{X}$, no substantial extra difficulties arise; we can immediately extend the results we have obtained concerning the estimation of $\bar{X}$.

The s.r. sample estimator of X_T which is commonly used is

$$x_T = N\bar{x},$$

and the earlier results confirm that x_T is unbiased for X_T; that is,

$$\boxed{E(x_T) = X_T}$$

and

$$\boxed{\operatorname{Var}(x_T) = (1-f)N^2S^2/n} \qquad (2.11)$$

Furthermore, x_T *is the minimum variance linear unbiased estimator of* X_T *based on a simple random sample of size n.*

With similar qualifications concerning the sample size, n, and value of the sampling fraction, f, we can use the normal approximation

$$x_T \sim N[X_T, (1-f)N^2S^2/n] \qquad (2.12)$$

to construct confidence intervals for X_T, or to choose a sample size to meet specified requirements concerning the precision of estimation of X_T.

For example, if n is more than about 40, an approximate $100(1-\alpha)\%$ *symmetric two-sided confidence interval* for X_T is given by

$$x_T - z_\alpha NS\sqrt{[(1-f)/n]} < X_T < x_T + z_\alpha NS\sqrt{[(1-f)/n]}.$$

For smaller n, the use of percentage points $t_{n-1}(\alpha)$ for the t-distribution with $(n-1)$ degrees of freedom, in place of z_α, is preferable.

Let us consider the question of choosing n to ensure that

$$Pr\{|x_T - X_T| > d\} \leq \alpha.$$

Using the normal approximation (2.12), this requires

$$n \geq N\left\{1 + \frac{1}{N}\left(\frac{d}{z_\alpha S}\right)^2\right\}^{-1}. \qquad (2.13)$$

Equivalently, we require

$$\operatorname{Var}(x_T) \leq (d/z_\alpha)^2 = V,$$

so that (2.13) becomes

$$n \geq \frac{N^2S^2}{V}\left[1 + \frac{1}{N}\frac{N^2S^2}{V}\right]^{-1}.$$

So if NS^2/V is very much less than 1, it will be reasonable to take

$$n_0 = \frac{N^2S^2}{V}$$

as the required sample size; otherwise we must use

$$n_0(1 + n_0/N)^{-1}.$$

In some respects it is more natural to express the accuracy we require of an estimator of X_T (or even of $\bar{X}$) in proportional, rather than absolute, terms. Thus we might ask what sample size is needed to ensure that

$$Pr\{|x_T - X_T| > \xi X_T\} \leqq \alpha. \qquad (2.14)$$

For example we may want to be at least 95% certain that x_T is within 2% of X_T. Then $\alpha = 0.05$, $\xi = 0.02$.

But this is less straightforward. It becomes necessary to replace d in (2.13) by ξX_T, and the appropriate value of n now depends on X_T, the unknown quantity we are trying to estimate. The best we can hope for is to obtain a rough estimate of the sample size necessary to satisfy (2.14), by replacing X_T in the right hand side of the inequality by some provisional estimate (possibly based on earlier surveys or experience).

Example 2.4

To obtain an early indication of the total sales of Christmas Cards throughout a network of 243 retail stationery shops, it is decided that a random sample of the shops should submit returns of their card sales by the end of January. How large a sample is needed to estimate total sales to within 10% of the correct figure with 95% assurance?

By July of each year precise figures of total sales of cards are available. For the previous three years the number of shops in the network has remained much the same; the total card sales, and standard deviations of sales from shop to shop, have been (in units of 10 000 cards)

X_T	S
321·7	0·826
366·8	0·776
401·0	0·804

So for the current year we might reasonably expect that X_T and S will be of the order of (say) 420 and 0·8, respectively. To obtain the required precision from the January returns it will consequently be necessary to take a simple random sample of size n, where, from (2.13),

$$n \leqq 243 \left\{1 + \frac{1}{243}\left(\frac{0.10 \times 420}{0.8 \times 1.96}\right)^2\right\}^{-1}$$

$$= 243\{1 + 2.96\}^{-1}$$

$$= 61.48.$$

So a sample of size $n = 62$ is needed.

Here $n_0 = N^2 S^2 / V = 82.30$, so that $n_0/N = 0.34$ and we do in fact need to use the more accurate expression $n_0(1+n_0/N)^{-1}$ to obtain the above value of 62 for the required sample size.

Suppose that such a sample of size 62 yields an estimate

$$x_T = 427.4,$$

then we obtain an approximate 95% confidence interval for X_T as

$$385.6 < X_T < 469.2,$$

reflecting (roughly) the 10% absolute accuracy we sought.

When attempting to estimate X_T with prescribed precision, we again typically encounter difficulties because we are unlikely to know the value of S^2, so that some preliminary indication of its value will be needed using one of the methods (i), (ii), (iii), or (iv) described in Section 2.5. Example 2.4 illustrates method (ii), in a case where the previous information arises from a complete enumeration, rather than a survey, of a similar situation.

Finally we should note that the population size N *needs to be known* if we are to use x_T to estimate X_T or to assess the sampling behaviour of x_T. This is true also for the study of $\bar{x}$ as an estimator of $\bar{X}$. In most situations N will be known, or can be estimated with fair accuracy. Where this is not so, difficulties arise even in the very choice of a s.r. sample.

2.7 Estimating a Proportion, P

Consider an engineering process in which a special component is produced for use in the assembly of a car. If some dimension, X, is not within required tolerances, the component will not be able to be used. To estimate the *proportion*, P, of useful components in a large batch, a s.r. sample of size n is taken and a count is made of the number, r, which have satisfactory values of X. The population of X values is not in itself of interest, we would like to know merely the proportion P of such values which lie within the tolerance limits.

Rather than studying a *quantitative* measure, X, in relation to whether it satisfies some criterion, we may on other occasions be concerned directly with some *qualitative* attribute or characteristic. What proportion P of the inhabitants of Renton live in rented accommodation? Again, a s.r. sample of size n gives an indication of P. If r out of the n

live in rented accommodation we might estimate P by the *s.r. sample proportion*,

$$p = r/n.$$

Once more we can readily modify the results for estimation of the *population mean*, $\bar{X}$, to describe the properties of the estimator p. Suppose P represents the proportion of members of a finite population of size N who possess some characteristic A. We define a variable Y_i describing the ith member of the population so that

$$Y_i = 1 \quad \text{if the member possesses characteristic A,}$$

$$= 0 \quad \text{otherwise.}$$

Then

$$Y_T = \sum_1^N Y_i = R$$

is the number of members possessing the characteristic A. Consequently,

$$\bar{Y} = \frac{1}{N}\sum_1^N Y_i = R/N = P,$$

so that the proportion P is merely the *population mean* for the Y values. Likewise, the sample proportion p is just the *mean* $\bar{y}$ for the *sample* of Y values. In discussing the performance of p as an estimator of P we are *once again considering the use of a s.r. sample mean to estimate the corresponding population mean*. The only essential difference arises from the simple structure of the population of Y values, where only the values 0 or 1 can occur. This implies a relationship between the population mean $\bar{Y}$ (or P) and the population variance S^2, which is now

$$S^2 = \frac{1}{N-1}\sum_{i=1}^N (Y_i-P)^2 = \frac{NP(1-P)}{N-1}, \qquad (2.15)$$

with a corresponding effect for the sampling behaviour of the estimator p. We will put $Q = 1-P$. Then from the results in Section 2.2 we have

$$\boxed{E(p) = P, \quad \text{so that } p \text{ is } unbiased \text{ for } P,}$$

and

$$\boxed{\text{Var}(p) = (1-f)S^2/n = \frac{(N-n)}{(N-1)}PQ/n.} \qquad (2.16)$$

But if P is unknown, S^2 will not be known. We can estimate S^2 by the unbiased estimator

$$s^2 = \frac{1}{(n-1)} \sum_1^n (y_i - \bar{y})^2 = npq/(n-1),$$

where $q = 1 - p$. Thus an *unbiased estimator* of $\mathrm{Var}(p)$ is given by

$$s^2(p) = (1-f)pq/(n-1).$$

Note that this is *not* the sample analogue $(N-n)pq/n(N-1)$ of (*2.16*) as we might intuitively think, although in practice the difference is unlikely to be important.

If the sampling fraction f is negligible, the estimator of $\mathrm{Var}(p)$ takes the simple form

$$s^2(p) = pq/(n-1).$$

This holds, in particular, when we are sampling from an infinite population.

2.8 Confidence Intervals for *P*

In sampling attributes or characteristics to estimate a proportion P, we know more about the sampling distribution of our estimator, p, than in the corresponding situations of estimating $\bar{X}$ or X_T. Indeed, the exact distribution of p is known! The number, r, of the sample members possessing the required attribute has a *hypergeometric distribution*,

$$p(r) = \frac{\binom{R}{r}\binom{N-R}{n-r}}{\binom{N}{n}} \; ; \quad \max(0, n-N+R) \leqq r \leqq \min(R, n).$$

We could thus make exact probability statements about r as a basis for constructing confidence intervals for P. Calculations are eased by using published tables or charts of cumulative probabilities for the hypergeometric distribution, particularly those especially designed for the purpose of making confidence statements about P. For example, Chung and DeLury (1950) give charts of 90%, 95%, and 99% confidence limits for P for $N = 500$, 2500, and 10000 for various values of n and p. Tables of cumulative probabilities for the hypergeometric distribution, for more modest values of N (up to 20), are given in Owen (1962). But such sources provide only a limited aid; it is not possible to obtain very accurate results from the charts; direct tabulations cover a relatively small range of values and can require complicated inverse interpolation. The knowledge of the exact distribution of *r* turns out to be rather unimportant *in practice* in view of the tedious calculations involved in the use of the hypergeometric distribution.

Consequently we must again seek useful approximations to the sampling distribution of the estimator, but now in a spirit of pragmatism rather than fundamental necessity. One obvious possibility is to use the *binomial distribution* as an approximation to the hypergeometric distribution. If *n* is small relative to both *R* and $(N-R)$, the lack of replacement of sampled members of the population can be ignored and *r* has essentially a binomial distribution, $B(n, P)$. We could use this binomial distribution to construct confidence intervals for *P*. But again this is of limited utility. The binomial distribution also involves tedious calculation (with inverse interpolation). Only if *n* is quite small, so that the calculations are reasonable, or published tables of confidence limits relevant (e.g. Fisher and Yates (1973)), is it worth proceeding with the binomial approximation.

In most applications we will find it convenient to go one stage further and use the *normal approximation to the binomial distribution*. Thus we will effectively assume that

$$p \sim N(P, (1-f)PQ/n) \qquad (2.17)$$

as a basis for constructing approximate confidence intervals for *P*. Notice how this is not the immediate extension of the argument supporting the binomial distribution for *p*, since some account is taken of the 'lack of replacement' in incorporating the f.p.c. in $\mathrm{Var}(p)$. In comparison with (*2.16*), we have omitted a factor $N/(N-1)$ from $\mathrm{Var}(p)$, which is justified for the sizes of population where the normal approximation will be used.

The normal approximation (*2.17*) will be reasonable, provided:

  (i) *n* is not too large relative to *R* or $(N-R)$,
 (ii) the smaller of nP and nQ is not too small; for example $\min(nP, nQ) > 30$ should suffice. If *P* is in the region of $\frac{1}{2}$, much smaller values of nP will be acceptable. Of course the values of nP and nQ will need to be assessed through their unbiased estimators np and nq.

We can determine confidence intervals for *P* from (*2.17*) in the usual manner of applying the normal approximation to the binomial distribution. Thus, from (*2.17*)

$$Pr\left\{\frac{|P-p|}{\sqrt{\left(\frac{(1-f)PQ}{n}\right)}} < z_\alpha\right\} = 1-\alpha, \qquad (2.18)$$

so that an approximate $100(1-\alpha)\%$ *two-sided confidence interval* for P is given as the region between the two roots of the quadratic equation

$$P^2[1+z_\alpha{}^2(1-f)/n]-P(2p+z_\alpha{}^2(1-f)/n]+p^2 = 0.$$

This can be simplified even further if n is sufficiently large. Replacing $Var(p)$ in (*2.17*) by its unbiased estimator $s^2(p)$ we have an approximate $100(1-\alpha)\%$ two-sided confidence interval for P with limits

$$p\pm[z_\alpha\sqrt{\{(1-f)pq/(n-1)\}}]. \qquad (2.19)$$

We could also introduce the usual continuity correction to take account of the fact that we are approximating a discrete distribution by a continuous one. This is particularly relevant when n is near the limit for justification of the normal approximation; it will tend to correct the length of the interval which would otherwise be too short.

The circumstances under which we may proceed to these various stages of approximation are not easily described in a concise manner. Some indication has been given in the discussion above. The principal determinants are the values of n and of NP and NQ relative to N. Some further details, with numerical illustration, are given by Cochran (1963, Chapter 3).

2.9 Choice of Sample Size in Estimating a Proportion

Consider the effect of the form (*2.16*) for $Var(p)$. Clearly this will be a maximum when $P = Q = \frac{1}{2}$, so that for a given sample size n we will be able to estimate P *least accurately* when it is in the region of $\frac{1}{2}$. This effect is more fully assessed by considering the value of $\sqrt{(PQ)}$ (reflecting the *standard error* of p). For $\frac{1}{4} < P < \frac{3}{4}$, $\sqrt{(PQ)}$ only varies over the range (0·433, 0·500), and little change occurs in the accuracy of the estimator p. P needs to be in the region of 0·07 (or 0·93) before the standard error is reduced to 50% of its maximum value.

Suppose we wish to estimate P with a s.r. sample large enough for the standard error (S.E.) of the estimator p to be no more than 2%. How large a sample would be needed? This will depend on what we mean by 'no more than 2%'. Is this a statement of the *absolute* value of the standard error, or do we require the standard error to be no more than 2% of P, i.e. are we concerned with the *relative* value of the standard error? The results will be quite different in the two cases!

(i) For an absolute value, we want

$$\text{S.E.}(p) = \sqrt{(PQ/n)} \leqq 0{\cdot}02$$

(assuming that the population is large so that the f.p.c. can be ignored).

(ii) For a relative value, we want

$$\text{S.E.}(p)/P = \sqrt{(Q/nP)} \leqq 0{\cdot}02.$$

For a selection of different values of P we obtain required sample sizes, n, as follows

P	0·02	0·04	0·10	0·20	0·40	0·50	0·60	0·80	0·90	0·96	0·98
(i) n	49	96	225	400	600	625	600	400	225	96	49
(ii) n	$1{\cdot}225 \times 10^5$	6×10^4	$2{\cdot}25 \times 10^4$	10^4	3750	2500	1667	625	278	105	52

Note how, in the case of prescribing the *relative* value of the standard error of p, the required sample size inevitably increases consistently with decreasing values of P and soon becomes unreasonably large for most practical purposes. This is in spite of the accuracy requirements (a 2% *coefficient of variation*) not being particularly stringent. And yet this is likely to be the type of requirement we would impose in situations where interest centres on estimating by Np the total number of individuals, $R = NP$, in the population who possess the defining characteristic. To keep the sample size manageable when P is small, we may well have to be content with a coefficient of variation somewhat larger than the 2% considered above.

The choice of a sample size to ensure certain limits on the standard error or coefficient of variation is of course equivalent to our earlier interest in achieving, with prescribed probability, a specified absolute, or proportional, accuracy for the estimator itself.

Thus to choose n to ensure that

$$\text{(i)} \qquad Pr\{|p-P| > d\} \leqq \alpha, \qquad\qquad (2.20)$$

or

$$\text{(ii)} \qquad Pr\{|p-P| > \xi P\} \leqq \alpha, \qquad\qquad (2.21)$$

amounts (using the normal approximation and ignoring the f.p.c.) to choosing n so that

(i)
$$\text{S.E.}(p) = \sqrt{\left(\frac{PQ}{n}\right)} \leq d/z_\alpha,$$

or

(ii)
$$\text{S.E.}(p)/P = \sqrt{(Q/nP)} \leq \xi/z_\alpha.$$

In practical situations we must again recognise that the standard error, or coefficient of variation, of p will not be known precisely since they depend on P, the quantity we are estimating. However, one facility we now have which was not present when estimating $\bar{X}$ or X_T is that we can place an upper bound on the sample size required to achieve a required *absolute* accuracy in the estimation of P, *whatever the value that P happens to have.*

To satisfy (*2.20*) we need

$$n > PQz_\alpha^2/d^2.$$

But PQ has a maximum value of $\frac{1}{4}$, when $P = \frac{1}{2}$. So that taking $n = z_\alpha^2/4d^2$ will certainly satisfy (*2.20*). Furthermore, this will not be too extravagant a policy over a quite wide range of values of P, say ($0\cdot30 < P < 0\cdot70$). No similar facility is available if we want to ensure a certain *proportional* accuracy—see tabulated values above.

Several loose ends remain to be tied up. It may be that the f.p.c. cannot be ignored (since the sampling fraction may need to be sizeable to ensure the required accuracy). Then again, we may be investigating a rather rare (or rather common) attribute in the population, so that the implication of assuming that $P = \frac{1}{2}$ in determining n to satisfy (*2.20*) will be to grossly *oversample* the population. Finally we may want to ensure a prescribed *proportional* accuracy.

Consider first of all the effect of retaining the f.p.c. and using the exact form (*2.16*) for $\text{Var}(p)$. To satisfy (*2.20*) we need (assuming the normal approximation to be justified)

$$n \geq N\left\{1 + \frac{(N-1)}{PQ}\left(\frac{d}{z_\alpha}\right)^2\right\}^{-1}$$

or, putting $(d/z_\alpha)^2 = V$,

$$n \geq \frac{PQ}{V}\left\{1 + \frac{1}{N}\left(\frac{PQ}{V} - 1\right)\right\}^{-1}. \tag{2.22}$$

So as a first approximation to the required sample size, we have

$$n_0 = \frac{PQ}{V},$$

which is just what is obtained above from ignoring the f.p.c.

If n_0/N is *not* negligible, then we must use the more exact expression (2.22) to obtain

$$n = n_0\{1 + (n_0 - 1)/N\}^{-1}.$$

Similarly, from (2.21), for a required relative accuracy, we need

$$n \geqq N \left\{ 1 + (N-1)\, \frac{\xi^2 P}{z_\alpha^2 Q} \right\}^{-1}$$

$$= \frac{Q}{P} \left(\frac{z_\alpha}{\xi} \right)^2 \left\{ 1 + \frac{1}{N} \left[\frac{Q}{P} \left(\frac{z_\alpha}{\xi} \right)^2 - 1 \right] \right\}^{-1}. \qquad (2.23)$$

Hence

$$n_0 = \frac{Q}{P} \left(\frac{z_\alpha}{\xi} \right)^2$$

(as obtained above for the required sample size when ignoring the f.p.c.), and

$$n = n_0\{1 + (n_0 - 1)/N\}^{-1}.$$

In conclusion we must take account of the fact that P will not be known precisely (otherwise the survey would be pointless), so that (2.22) and (2.23) are not directly applicable. Again the methods (i), (ii), (iii), and (iv) of Section 2.5 must be considered as means of providing some 'advance estimate' of P for the purpose of determining the required sample size. A pilot study will yield a preliminary estimate of P—this essentially combines methods (i) and (iv). The upper bound, provided by assuming that $P = \frac{1}{2}$ when *absolute* accuracy is of interest, also illustrates the use of method (iv). Again, and this is by no means uncommon, allied surveys (at an earlier time or on a related topic) may provide a reasonable idea of the value of P, as may published statistics on a wider front. For example, suppose a large nationally based survey was conducted two years ago to investigate family wealth. It showed that 9% of families possessed more than 1 car. If we decide to estimate the same measure on some particular local community we might choose to act as if P is somewhere in the region of 0·09 in order to determine a required sample size. Needless to say we must ensure that the concept of the 'family' is a similar one in the two cases, and take note of any particular social structure in the local survey which may distinguish it from the national one (the two year time separation is also relevant, of course).

Example 2.5

Suppose, in the context of Example 2.3, that a somewhat liberal view is taken of 'casual holidays', recognising that the work is of such a nature that workers will feel the need to 'take the odd day off on the spur of the moment'. Up to 3 days in six months is regarded as reasonable; we want to estimate the proportion of workers taking more than 3 days off in the six months of study. From the figures given in Example 2.3 we find

$$p = 88/1000 = 0.088,$$

and obtain a 95% confidence interval for P, from (2.19), as

$$0.071 < P < 0.105$$

(Introducing the continuity correction yields $0.070 < P < 0.106$)

The more accurate form obtained from solving the quadratic equation is

$$0.072 < P < 0.107.$$

These results still show some minor discrepancies. But these are hardly of any practical importance. Even ignoring the f.p.c. in (2.19) gives a very similar interval: $0.0704 < P < 0.106$.

2.10 Extensions

Sub-populations

In the industrial example on 'casual holidays' (Examples 2.3 and 2.5), it is likely that quite different patterns of behaviour exist for different groups of workers. Different average numbers, and total numbers, of days 'casual holiday' may have been taken by the workers in the *sub-populations* for these different groups of workers. The population we have examined in Examples 2.3 and 2.5 is the combination of all the sub-populations; our estimates or inferences relate to this aggregate. But we may well wish to study the characteristics of the sub-populations separately. What group of workers takes the most, or least, number of days 'casual holiday'? What is the behaviour pattern for some particular group?

Thus if there are k sub-populations with means $\bar{X}_i$, totals X_{iT}, variances S_i^2, or proportions $P_i (i = 1, \dots, k)$, we may wish to estimate these characteristics separately. If we take advance note of this interest and can take simple random samples from each sub-population separately, the methods described above and their corresponding properties will apply to each sub-population. But suppose we have taken a simple random sample from the aggregate population. How are we now to study the sub-populations? Subsequent assignment of observations to the sub-populations *does not yield* s.r. samples in these sub-populations —consequently the earlier results do not apply. The problem is that the sample sizes are not predetermined, but are themselves *random* quantities subject only to a constraint on their sum. More detailed study is now needed of how to analyse the data appropriately, in order to estimate the $\bar{X}_i$, X_{iT}, S_i^2, or P_i. Some of the results of such study are described by Cochran (1963, Chapter 2).

Such structured populations present a complementary possibility. It may be that in recognising the different natures of the sub-populations, and sampling these sub-populations separately, we might be able to increase the efficiency of our study of the aggregate population. This is indeed true under certain circumstances, as we shall see later (Chapter 4) when we consider the idea of *stratification*.

Multiple Attributes

We have considered at some length the estimation of a proportion P, of a population, falling into some category. The classification of members of the population was a dichotomous one—each member was either in the category of interest, or not. Often a more complex classification exists, and population members must be assigned to one of *several* categories. For example, in an enquiry into social structure we may want to estimate the proportions of people in different social classes, or in different nationality groups. Or again in the industrial example above, we might wish to estimate the proportions of workers in the different occupational groups.

Thus instead of a single proportion P, we have proportions $P_1, P_2, \dots$ that must be estimated from the appropriate assignment of the individuals in a simple random sample. The numbers, $n_1, n_2, \dots$, in the different categories now follow the natural extension of the hyper-geometric distribution of Section 2.8, but again if the n_i are small in relation to the population totals $N_1, N_2, \ldots$ we can reasonably approximate this distribution. In this case we would use the *multinomial distribution*

$$\frac{n!}{n_1! \, n_2! \ldots} \, P_1{}^{n_1} \, P_2{}^{n_2} \ldots$$

where previously we used the binomial distribution, as the basis for drawing inferences about $P_1, P_2, \ldots$.

Exercises for Chapter 2

I

Two independent s.r. samples of sizes 200 and 450 were chosen one after the other (without replacement) from a population of 2400 students in a non-residential College. Each student was asked the distance (in miles) from the College that he or she lived. The sample means and variances were

$$\bar{x}_1 = 5 \cdot 14, \qquad \bar{x}_2 = 4 \cdot 90,$$
$$s_1{}^2 = 3 \cdot 87, \qquad s_2{}^2 = 4 \cdot 02.$$

Calculate an approximate 99 % confidence interval for the mean distance from the College that students live.

II

Consider the 'casual holidays' problem described in Examples 2.3 and 2.5. From company computer records it might be easy to obtain precise information on the number of workmen who missed no workdays over the six months period of interest. Suppose that $45 \cdot 82 \%$ of the workforce of 36 000 missed no work. A s.r. sample of 500 men out of the remainder yielded the following results.

'Days off'	1	2	3	4	5	6	7	8	9	10
No. of men	157	192	90	31	18	5	2	4	0	1

Estimate the total number of days 'casual holiday' taken over the six month period; and determine the approximate standard error of the estimator. Do the same calculations for the s.r. sample of size 1000 described in Example 2.3, and explain the discrepancies in the approximate standard errors in the two cases.

III

In a private library the books are kept on 130 shelves of similar size. The numbers of books on 15 shelves picked at random were found to be

28, 23, 25, 33, 31, 18, 22, 29, 30, 22, 26, 20, 21, 28, 25.

Estimate the total number, X_T, of books in the library, and calculate an approximate 95% confidence interval for X_T.

Suppose the resulting estimate is not accurate enough; we want to be 95% sure that a s.r. sample estimate of X_T is within 100 of the true value. How many shelves should be included in the sample?

IV

A s.r. sample of size $2n$ is chosen from a finite population of size $N (N > 2n)$. The population mean and variance are $\bar{X}$ and S^2, respectively. The sample is divided into two equal parts: the first n observations, and the second n observations. The sub-sample means are $\bar{x}_1$ and $\bar{x}_2$. Derive a simple unbiased estimator of S^2 based on n, $\bar{x}_1$, and $\bar{x}_2$.

V

A residential area has 5000 private houses. We want to estimate the proportions of houses with

(a) more than three persons living in them.
(b) more than one car owned by the occupants of the house.

The estimators are required to have standard errors not exceeding 0·02 and 0·01, respectively. From other surveys it would appear that the proportions, for (a) and (b), will lie in the ranges 0·35 to 0·55 and 0·10 to 0·20, respectively. The two proportions are to be estimated from a single s.r. sample. How large a sample is needed to meet the accuracy requirements?

3 Ratios: Ratio and Regression Estimators

We have so far considered the problems associated with the estimation of a single population characteristic, based on the probability sampling scheme of simple random sampling. Continuing to study the same sampling scheme, we shall now broaden our enquiries a little with respect to the population characteristics of interest. Frequently (indeed predominantly) the aim of a sample survey is to seek information simultaneously on a range of different measures in the finite population we are studying. Both our practical interest and the cost and effort of conducting a survey demand that we should do so.

For example, in the sociology example (a) in Chapter 1, there will be a variety of different aspects of the attitudes of 18 year olds to recent legislative changes that will be of interest concerning their psychological make-up in relation to new liberties or restrictions presented to them. Responses are likely to consist of complete questionnaires concerning such matters, and also providing descriptive details of the respondents as a guide to the representativeness of the sample. Initial contact with the respondents is the major cost and administration factor. To restrict attention to a single question ('Do you feel that the right to vote gives you an important say in the organisation of society?') is both sterile and inefficient. (Note also that this example is a leading question which can prejudice the reaction of the respondent!) Attitudes are complex, involving many different aspects, and may differ widely with the personal circumstances of the individual. To seek answers to, say, 20 questions involves little more trouble than obtaining the answer to one question; it provides far greater facility for assessing attitudinal factors and can yield tangential information on the population for current or future use.

Thus we are often confronted with multivariate data concerning a variety of measures in the population, represented by variables X, Y, Z, ... Simultaneous estimation of population characteristics exploiting the correlation structure of the multivariate population is not a part of this introductory treatment. However, one simple extension of the univariate situation will be considered in detail in this chapter. This concerns the *bivariate* case where we simultaneously observe two variables, X and Y. We shall discuss two matters

(i) how to estimate the ratio of two population characteristics, for example $\bar{X}/\bar{Y}$,
(ii) how simultaneous observation of X and Y, exploiting any association between these variables, can in certain circumstances assist in the efficient estimation of the characteristics of *one* of them, for example $\bar{X}$.

3.1 Estimating a Ratio

In a variety of situations we may need to estimate a ratio of two population characteristics: the means, or totals, of two variables X and Y. We will be interested in the quantity

$$R = \bar{X}/\bar{Y} = X_T/Y_T,$$

the *population ratio*.

This interest can arise in two ways. Either the ratio is of intrinsic interest in its own right. For example, we may wish to estimate the proportion of arable land given over to the growth of barley in some geographic region. To this end we might sample farms in the region and record their total acreage, and the acreage of barley crops. If these are Y_i and X_i for the different farms in the region, it is precisely $R = X_T/Y_T$ that we must estimate. Alternatively, concern for the ratio R may be forced on us by administrative convenience in the construction of a viable sampling scheme. Suppose we want to estimate the average annual income per head, or average number of cars per person, for adult persons living in a metropolitan district. We might envisage taking a simple random sample of adult individuals, noting their income or the numbers of cars they possess (predominantly 0 or 1) and using the sample mean in each case to estimate the corresponding population mean of interest. But it is unlikely to be easy to sample adult individuals at random—ease of access to the population, and other quantities of interest in their own right, will often lead to the use of larger sampling units, say households. If this is the case, we become concerned with ratios, rather than means. The average income per head is now best regarded as the ratio of total income X_T to the total adult population size Y_T, with both characteristics estimated from the sample of households. Similarly for the car-ownership enquiry.

Note two features of this example: the use of groups of individuals (as sampling units) in the study of characteristics *per individual*; also the simple nature of one of the variables in being an indicator variable taking just the values 0 or 1. Both features commonly arise in ratio estimation (although the barley example shows that we need not always encounter a simple indicator variable).

Thus we wish to estimate the population ratio $R = X_T/Y_T$, on the basis of a simple random sample $(x_1, y_1), \ldots, (x_n, y_n)$ of the bivariate population measures (X_i, Y_i), $(i = 1, \ldots, N)$.

The estimator predominantly used is the corresponding *sample ratio*

$$r = \bar{x}/\bar{y} = x_T/y_T.$$

Since both numerator and denominator reflect random variation, the distribution of r is complicated in form. It is likely to be skew in small samples, and r turns out to be somewhat biased. In large samples the bias becomes negligible; the distribution of r tends to normality, and inferences may now be based on the normal distribution with appropriate variance, $\mathrm{Var}(r)$.

We must consider this in more detail. In large samples we have, approximately,

$$\boxed{E(r) = \bar{X}/\bar{Y} = X_T/Y_T}$$

and

$$\boxed{\mathrm{Var}(r) = \frac{1-f}{n\,\bar{Y}^2} \sum_{i=1}^{N} \frac{(X_i - RY_i)^2}{N-1}} \qquad (3.1)$$

where f is the sampling fraction n/N.

These approximate results may be obtained by writing

$$r - R = \bar{x}/\bar{y} - R = \frac{\bar{x} - R\bar{y}}{\bar{y}}$$

and replacing $\bar{y}$ in the denominator by $\bar{Y}$, which should be reasonable in large samples in view of the *consistency* of the s.r. sample mean. We obtain

$$E(r - R) = \frac{E(\bar{x}) - RE(\bar{y})}{\bar{Y}} = \frac{\bar{X} - R\bar{Y}}{\bar{Y}} = 0.$$

Also, on this approximation,

$$\text{Var}(r) = E[(r - R)^2] = \frac{1}{\bar{Y}^2} E[(\bar{x} - R\bar{y})^2].$$

But if we define

$$Z_i = X_i - RY_i,$$

then $(\bar{x} - R\bar{y})$ is just the mean, $\bar{z}$, of a simple random sample of size n chosen from the population of Z_i values.

Since this derived population has *zero* mean, we have from (2.3)

$$\text{Var}(r) = (1 - f)S_z^2/(n\bar{Y}^2),$$

where S_z^2 is the variance of the population of Z-values.

Thus,

$$\text{Var}(r) = \frac{1 - f}{n\bar{Y}^2} \sum_{i=1}^{N} Z_i^2/(N - 1)$$

$$= \frac{1 - f}{n\bar{Y}^2} \sum_{i=1}^{N} \frac{(X_i - RY_i)^2}{N - 1}.$$

Equivalently we can make use of standard results on the asymptotic form of the mean and variance of the ratio of two statistics. We have that

$$E\left(\frac{\bar{x}}{\bar{y}}\right) = \frac{E(\bar{x})}{E(\bar{y})} + 0\,(n^{-1}), \tag{3.2}$$

and

$$\text{Var}\left(\frac{\bar{x}}{\bar{y}}\right) = \frac{\text{Var}(\bar{x})}{[E(\bar{y})]^2} - \frac{2E(\bar{x})}{[E(\bar{y})]^3} \text{Cov}(\bar{x}, \bar{y}) + \frac{[E(\bar{x})]^2}{[E(\bar{y})]^4} \text{Var}(\bar{y})$$

$$+ 0\,(n^{-\frac{3}{2}}). \tag{3.3}$$

But

$$E(\bar{x}) = \bar{X}, \; E(\bar{y}) = \bar{Y},$$

$$\text{Var}(\bar{x}) = \frac{(1 - f)}{n} S_X^2,$$

$$\text{Var}(\bar{y}) = \frac{(1 - f)}{n} S_Y^2,$$

so that from (3.2) we again demonstrate that r is approximately unbiased in large samples. The bias in r is seen to be of order n^{-1}. From the leading term in (*3.3*) we can again obtain the approximation (*3.1*), using the fact that

$$\text{Cov}(\bar{x}, \bar{y}) = \frac{1-f}{n}\text{Cov}(X, Y) = \frac{(1-f)S_{XY}}{n},$$

where the covariance, S_{XY}, of the bivariate population of (X_i, Y_i) values is *defined* as

$$S_{XY} = \frac{1}{N-1}\sum_{1}^{N}(X_i - \bar{X})(Y_i - \bar{Y}). \tag{3.4}$$

The variance of r is of order n^{-1} as we should expect, but we note that, in general, the approximation is correct to order $n^{-\frac{3}{2}}$. If the bivariate finite population manifests a roughly *normal* form, the error term in (*3.1*) is of order n^{-2} (in line with results for infinite bivariate normal populations), and the approximation for $\text{Var}(r)$ is correspondingly more accurate.

But (*3.3*) also yields an alternative form for (*3.1*), namely the approximation

$$\text{Var}(r) = \frac{1-f}{n\bar{Y}^2}\{S_X^2 - 2RS_{XY} + R^2 S_Y^2\}, \tag{3.5}$$

which explicitly includes the population covariance, S_{XY}.

Further details on the extent of the bias in r and on the adequacy of the approximate form for its variance are given in Cochran (1963, Chapter 6).

Once again we encounter the familiar problem that the variance of our estimator is expressed in terms of *population* characteristics, which will be unknown. Thus we will need to *estimate* $\text{Var}(r)$ from our data, and it is usual to employ the direct sample analogue

$$s^2(r) = \frac{(1-f)}{n\bar{y}^2}\sum_{i=1}^{n}\frac{(x_i - ry_i)^2}{n-1}.$$

This differs from (*3.1*) by a term of order n^{-2}.

The sum of squares

$$\sum_{i=1}^{n}(x_i - ry_i)^2$$

is most conveniently calculated as

$$\sum_{i=1}^{n}x_i^2 - 2r\sum_{i=1}^{n}x_iy_i + r^2\sum_{i=1}^{n}y_i^2,$$

echoing the alternative form (3.5). Note how it is unnecessary to correct the x_i and y_i values for their sample means, owing to compensation effects.

We remarked above that the exact distribution of r is most complicated, but that it approaches normality in large samples (from large populations). Thus, for large samples, we can construct confidence intervals for R. If $s(r)$ is the sample estimate of the standard error of r, $[\mathrm{Var}(r)]^{\frac{1}{2}}$, we have an approximate $100(1-\alpha)\%$ *symmetric two-sided confidence interval* for R in the form

$$r - z_\alpha s(r) < R < r + z_\alpha s(r).$$

Example 3.1

A daily newspaper conducts a survey of food costs by taking a simple random sample of 48 basic foodstuffs purchased in a large supermarket. Prices (in pence) for these items are recorded on two separate occasions, three months apart, the earlier ones being denoted y_i, the later x_i. The sample ratio, $r = \bar{x}/\bar{y}$ gives an indication of the change in basic food prices over the three months period in the form of an estimate of the population ratio R of the mean prices of all foodstuffs on the two occasions.

The following results were obtained:

$$\bar{x} = 12{\cdot}07,\ \bar{y} = 11{\cdot}41;$$

$$\sum x_i^2 = 9270{\cdot}6,\ \sum y_i^2 = 8431{\cdot}7,\ \sum x_i y_i = 8564{\cdot}1.$$

Clearly the population size will be vast in relation to the sample size $n = 48$, so that we can ignore the f.p.c.

So the estimated ratio is $1{\cdot}06$: a 6% rise in prices over the three months in question. The approximate sampling variance of the estimator is

$$\frac{587{\cdot}0}{48 \times 47 \times (11{\cdot}41)^2} = (0{\cdot}0447)^2$$

so that we have an approximate 95% confidence interval for R as $0{\cdot}970 < R < 1{\cdot}145$. Clearly any firm statement of an average food price *increase* would be unwise in the light of this approximate interval (the wide range reflects the small sample size in the survey).

An intuitively obvious alternative estimator of R based on a s.r. sample $(x_1, y_1), \ldots, (x_n, y_n)$ is the sample average of the individual ratios x_i/y_i; that is,

$$\frac{1}{n} \sum_{i=1}^{n} (x_i/y_i).$$

This is not a very satisfactory estimator. It can be seriously biased and can have large E.M.S.E. (although it is possible to construct a modified estimator corrected for bias—see Cochran (1963, Section 6.15); also Chapter 6 later). See Section 5.3 for a discussion of a similar estimator in *cluster sampling*.

3.2 Ratio Estimator of a Population Total or Mean

Suppose we wished to estimate the total local government expenditure in 1971 on some particular service (health or education, say; let us in fact choose the provision of recreational facilities for children). We might decide to do this by sampling the different county and metropolitan authorities throughout the country and making specific enquiries on such expenditure in a simple random sample of these local authorities. Clearly there is going to be large variation in the amounts spent on recreational facilities by the different authorities. This will reflect many factors, including their geographic area, number of inhabitants, available budgetary resources, and rural, urban, and industrial breakdowns. We may have available a deal of information on these relevant factors in the population. If so, it would surely be desirable to make use of our knowledge of the structure of the population to assess the representativeness of any random sample we may draw, or to guide the choice of the sample in an attempt to obtain a more efficient estimator. We shall be considering at length in the next chapter the concept of *stratification* (the division of the population into non-overlapping groups, or *strata*, which represent its structure), and its use in the construction of what we hope will be better estimators than would be obtained from a simple random sample from the unstratified population at large.

At this stage, however, we will consider an alternative means of exploiting known elements of the population structure in certain circumstances. This consists of the use of ancillary quantitative information to construct what is called the *ratio estimate* of the population total (or mean). It seems reasonable, in the local authority example above, that expenditure on recreational facilities for children should change from one authority to another *roughly in proportion* to their number of inhabitants, or their total annual budgets. (Some slight anomalies may be observed in particularly small or large authorities, or for largely rural or industrial ones, but the pattern of proportionality overall is likely to appear quite strong.)

Suppose X_i denotes expenditure on recreational facilities for authority i, Y_i its number of inhabitants, and we sample both measures simultaneously, at random from the whole population, to obtain a s.r. sample of size n: $(x_1, y_1), \ldots, (x_n, y_n)$. The total number of inhabitants for the whole population, Y_T, is likely to be known fairly accurately (for example, from census returns); we will also know the number, N, of local authorities in the population. But we could have *estimated* Y_T from the sample by means of the estimator

$$y_T = N\bar{y},$$

where $\bar{y}$ is the s.r. sample mean. Similarly we could estimate the total expenditure X_T (the characteristic of principal interest) by

$$x_T = N\bar{x}.$$

The estimate y_T has no interest in its own right (since we know Y_T), but it has the important advantage that by comparing it with the population characteristic Y_T we can informally assess the representativeness of the sample. If y_T is very much less than Y_T, then in view of the rough proportionality of X_i and Y_i we would conclude that x_T is likely to underestimate X_T; if y_T is too large, so is x_T likely to be. If the proportionality relationship were *exact* we would have

$$X_i = RY_i \ (i = 1, \ldots, N), \tag{3.6}$$

where R is the population ratio, X_T/Y_T or $\bar{X}/\bar{Y}$, discussed in the previous section. Thus,

$$X_T = RY_T,$$

and we could estimate X_T by replacing R with the sample estimate, r, to obtain an estimate of the population total, X_T, in the form

$$x_{TR} = rY_T = \frac{Y_T}{y_T} x_T. \tag{3.7}$$

C

The estimator x_{TR} is called the s.r. sample *ratio estimator of the population total*. Note that this achieves precisely the type of compensation we require for values of y_T which are fortuitously larger, or smaller, than the known value, Y_T; it reduces, or increases, our estimate of X_T accordingly.

The *exact* case is discussed simply to motivate the estimator (*3.7*)—if (*3.6*) held, one observation (x_1, y_1) would determine R precisely and hence $X_T (= R Y_T)$, so that 'estimation' of X_T is trivial.

If the exact relationship (*3.6*) does not hold (it is hardly likely to do so in any practical situation), the same aim at compensation must still be sensible whenever there is 'rough proportionality' between the variable of interest, X, and the ancillary (concomitant) variable , Y. In such cases we can again use the ratio estimator (*3.7*).

If interest centres on the population mean $\bar{X}$, rather than the total X_T, then similar arguments support the use of the *ratio estimator of the population mean*,

$$\bar{x}_R = r\bar{Y} = \frac{Y_T}{y_T}\bar{x}. \tag{3.8}$$

Such ratio estimators have an obvious appeal, but clearly we must attempt to identify the circumstances under which we obtain an important improvement in efficiency of estimation over the direct s.r. sample total, x_T, or mean, $\bar{x}$. This must involve a clearer statement of what is meant by 'rough proportionality'.

A statistically important factor in these enquiries is that the sole sample statistic that is used is the *sample ratio*, r, whose properties have been discussed in some detail in the previous section.

Consider the estimator $\bar{x}_R$. From (*3.2*), $\bar{x}_R$ is *asymptotically unbiased*; in certain circumstances it is unbiased for all sample sizes, as we shall see shortly.

From (*3.1*) and (*3.5*) we see that the approximate variance of $\bar{x}_R$ (for large samples) is

$$\text{Var}(\bar{x}_R) = \frac{1-f}{n} \sum_{i=1}^{N} \frac{(X_i - RY_i)^2}{N-1} \tag{3.9}$$

$$= \frac{1-f}{n}(S_X^2 - 2RS_{XY} + R^2 S_Y^2)$$

$$= \frac{1-f}{n}(S_X^2 - 2R\rho_{XY}S_X S_Y + R^2 S_Y^2), \tag{3.10}$$

where

$$\rho_{XY} = \frac{S_{XY}}{S_X S_Y}$$

is defined to be the *population correlation coefficient*. If the exact relationship (3.6) held, then $\text{Var}(\bar{x}_R)$ would of course be zero (by (3.9)). In practice this will not be so, but $\text{Var}(\bar{x}_R)$ is clearly going to become smaller, the larger the (positive) correlation between X and Y in the population.

For estimating X_T we have analogous results for x_{TR}. It is asymptotically unbiased, and has large sample variance

$$\frac{N^2(1-f)}{n} \sum_{i=1}^{N} \frac{(X_i - RY_i)^2}{N-1}$$

or

$$\frac{N^2(1-f)}{n} (S_X{}^2 - 2R\rho_{XY}S_X S_Y + R^2 S_Y{}^2).$$

Again we will need to estimate $\text{Var}(\bar{x}_R)$, or $\text{Var}(x_{TR})$, from the sample, and the most convenient forms to use are

$$\frac{1-f}{n(n-1)} \left(\sum_{i=1}^{n} x_i{}^2 - 2r \sum_{i=1}^{n} x_i y_i + r^2 \sum_{i=1}^{n} y_i{}^2 \right),$$

and

$$\frac{(1-f)N^2}{n(n-1)} \left(\sum_{i=1}^{n} x_i{}^2 - 2r \sum_{i=1}^{n} x_i y_i + r^2 \sum_{i=1}^{n} y_i{}^2 \right),$$

respectively. Note that this estimation stage must introduce further inaccuracies; our only safeguard lies in the size of the sample.

Using the large sample forms for variances, and exploiting the asymptotic normality of the estimators, we can again obtain approximate confidence intervals for $\bar{X}$ or $\bar{X}_T$ in the usual way. A reasonable practical prescription for using the normal distribution and the approximate form for the variance seems to be that the sample size should be about 40, the sampling fraction no greater than 0·25, and that the ratios $S_X/\bar{X}$ and $S_Y/\bar{Y}$ should both be less than 0·10. These latter quantities are known as the population *coefficients of variation* for the X and Y variables. We shall denote them by C_X and C_Y, respectively.

When the large sample results are inappropriate, the assessment of the properties of $\bar{x}_R$ and x_{TR}, and the construction of confidence intervals for $\bar{X}$ and X_T using ratio estimators, are most complicated. The exact results are incompletely known, and not very tractable. Some more appropriate approximate results, which take account of the fact that the distribution of r frequently has positive skewness, are summarised by Cochran (1963, Chapter 6).

Let us return briefly to the question of the bias of the ratio estimators. The model (3.6) for the relationship between the X_i and Y_i values in the population was of little relevance apart from motivating the form of $\bar{x}_R$ or x_{TR}. We cannot expect to encounter it in practice; indeed if we did, there would be no estimation problem! Relaxing the model slightly we might consider one for which

$$X_i = RY_i + E_i, \tag{3.11}$$

with $\Sigma_y E_i = 0$, where Σ_y denotes summation over all subscripts i for which $Y_i = y$. (This is the finite population analogue of the classical linear regression model.)

In this case $\bar{X} = R\bar{Y}$ (as required from the definition of R), and in a s.r. sample of size n,

$$r = \frac{\bar{x}}{\bar{y}} = R + \frac{\bar{e}}{\bar{y}},$$

where $\bar{e}$ is the sample mean of the E values in the sample.

Clearly the conditional expectation

$$E(\bar{e}|y_1,\ldots,y_n) = 0$$

for this model, so that $E(r) = R$ and we conclude that *r is unbiased for all sample sizes*. But whilst more plausible than (3.6), the model (3.11) is again unlikely to be exactly satisfied by our finite population. At best we may find that the population is 'roughly' of this form, and may consequently be less concerned than otherwise about possible bias in $r, \bar{x}_R,$ or x_{TR}.

We have examined in some detail the properties of ratio estimators of a population mean or total, but a crucial question remains. Under what circumstances, if any, should we use a ratio estimator in preference to a s.r. sample mean or total? More effort is involved in obtaining $\bar{x}_R$ (or x_{TR}) than $\bar{x}$ (or x_T), albeit to only a slight degree since the major task is in designing and conducting the survey. Simultaneous measurement of two quantities, X and Y, often poses little more work than measurement of one alone. We can thus rule out differential costs in most circumstances. The prime consideration becomes the precision of the estimation principle; is $\bar{x}_R$ (or x_{TR}) more or less efficient than $\bar{x}$ (or x_T). The answer turns out to be that either possibility can arise, depending on the population correlation ρ_{XY} and the coefficients of variation C_X and C_Y.

We must identify the conditions under which $\text{Var}(\bar{x}_R)$ is less than $\text{Var}(\bar{x})$; that is where *the ratio estimator is the more efficient.* From (*2.3*) and (*3.10*) we see that

$$\text{Var}(\bar{x}_R) < \text{Var}(\bar{x})$$

if

$$R^2 S_Y^2 < 2R\rho_{XY}S_X S_Y,$$

that is, *if*

$$\rho_{XY} > \tfrac{1}{2}\frac{C_Y}{C_X} \qquad (3.12)$$

So we see that a gain in efficiency is not in fact guaranteed; we need the population correlation coefficient to be sufficiently large. (In practice we would need to assess the criterion (*3.12*) from sample estimates of ρ_{XY}, C_X, and C_Y.)

But notice that however large ρ_{XY} turns out to be, we still *need not necessarily* obtain a more efficient estimator by using $\bar{x}_R$ (or x_{TR}). *If*

$$C_Y > 2C_X, \qquad (3.13)$$

the ratio estimator x_R (or x_{TR}) *cannot possibly be more efficient than* $\bar{x}$ (or x_T) even with essentially perfect correlation between the X and Y values. Thus two factors are important for efficiency improvement from ratio estimators: the variability of the auxiliary variable Y must not be substantially greater than that of X (in the sense of (*3.13*)), and the correlation coefficient ρ_{XY} must be large and positive.

Nonetheless, many practical situations are encountered where the appropriate conditions hold and ratio estimators offer substantial improvement over $\bar{x}$ or x_T. Reviewing the situation we need the following circumstances to hold:

(i) We must be able to observe simultaneously two variables X and Y which appear to be roughly proportional to each other (that is, which have high positive correlation).

(ii) The auxiliary variable Y must not have a substantially greater coefficient of variation than X.

(iii) The population mean $\bar{Y}$, or total Y_T, must be known exactly.

The 'rough proportionality' in (i) implies a more or less linear relationship through the origin. The fact that this is through the origin has not been formally considered above. Its major importance is a negative one. If X and Y were essentially linearly related, but the relationship did *not* pass through the origin, then we might be well advised to consider the alternative so-called *regression estimator* discussed in the next section to see if it is better than either the ratio estimator, or the sample mean.

Example 3.2

Let us reconsider the Class Example data given in Section 1.6. For this population, with X denoting height and Y denoting weight, we clearly have quite a strong positive association between the X_i and Y_i values, as shown in the scatter diagram, Figure 3. Using our privileged knowledge of the

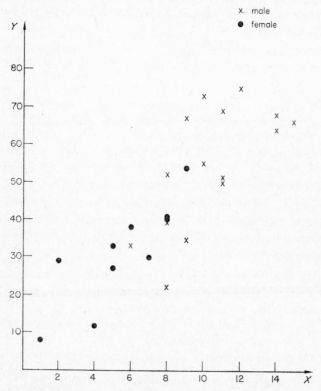

Figure 3   Scatter diagram of heights (X) and weights (Y) in the *Class Example*.

whole population we can calculate various population characteristics of interest.

$$S_X{}^2 = 12\cdot42, \quad S_{XY} = 56\cdot18, \quad S_Y{}^2 = 361\cdot19,$$

$$\rho_{XY} = 0\cdot839,$$

$$R = \bar{X}/\bar{Y} = 0\cdot1866, \quad C_X = 0\cdot418, \quad C_Y = 0\cdot420.$$

The ratio of the coefficients of variation is almost exactly unity; the correlation coefficient very much in excess of half this value. Although the results above refer to large samples we should perhaps not be too surprised to find that the ratio estimator of $\bar{X}$, based on a s.r. sample of size 5, greatly improves on the s.r. sample mean. To investigate this, 500 such s.r. samples have been generated and $\bar{x}_R$ evaluated in each case. Figure 4 shows a histogram of the values obtained. By comparison with Figure 2, we see that $\bar{x}_R$ is far less disperse than $\bar{x}$; consequently more efficient. The mean and variance of the 500 values of $\bar{x}_R$ are 8·45 and 0·69 respectively, in comparison with 8·46 and 1·94 for the 500 values of $\bar{x}$. (The large sample approximation to $\mathrm{Var}(\bar{x}_R)$ is 0·65.)

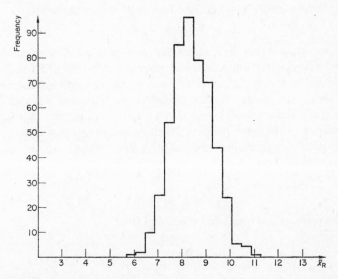

Figure 4   Histogram of 500 values of $\bar{x}_R$ from s.r. samples of size 5 in the *Class Example*.

3.3 Regression Estimator of a Population Total or Mean

Another type of estimator which aims at exploiting the relationship between some variable of interest, X, and an auxiliary variable, Y, in order to obtain greater precision in estimating $\bar{X}$ or X_T is the so-called *regression estimator*. This is a particularly useful estimator when (again) there is some degree of linearity in the relationship between the X and Y values in the population, but this relationship does not necessarily pass through the origin. The extreme case would be when

$$X_i = \bar{X} + B(Y_i - \bar{Y}) \qquad (3.14)$$

for all population values (X_i, Y_i) and some appropriate value of B. The regression estimator, like the ratio estimator, is applied in situations where the value of $\bar{Y}$ is known. In the case of a population satisfying (3.14), we could clearly determine $\bar{X}$ exactly from a single observation (x, y), since

$$\bar{X} = x + B(\bar{Y} - y).$$

But of course no such precise structure is likely to be encountered in real life problems. What is possible, however, is that the X and Y values do seem on inspection to vary in a way which reflects a degree of linearity, with relatively small superimposed deviations about the linear relationship. Thus, for example, we might conveniently adopt the model

$$X_i = \bar{X} + B(Y_i - \bar{Y}) + E_i, \qquad (3.15)$$

on the assumption that the E values have zero population mean and bear no systematic relationship to the Y values, and in the belief that the population variance, S_E^2, of the E_i will be rather small in relation to S_X^2. The model (3.15) is then a useful representation of a population where the variation in X values may be attributed in part to a linear dependence on the corresponding Y values, and in (perhaps lesser) part to population vagaries unconnected with the Y values. If we assume (as suggested above) that $S_{YE} = 0$, then we see how S_X^2 has two components,

$$S_X^2 = B^2 S_Y^2 + S_E^2,$$

and the population correlation coefficient is

$$\rho_{XY} = \frac{BS_Y}{S_X}.$$

So,

$$S_E^2 = S_X^2(1 - \rho_{XY}^2),$$

and we note that the relative importance of the Y and E values in accounting for the variability in the X values depends on the value of ρ_{XY}^2. If the X and Y values are highly correlated (in a positive or negative sense), the E values make little contribution, and vice versa. This mirrors the characteristics of the classical linear regression model in the infinite population case.

Suppose now that we draw a s.r. sample $(x_1, y_1), \ldots, (x_n, y_n)$ and, knowing $\bar{Y}$, want to estimate $\bar{X}$. It seems sensible that we should take account of any linear relationship in the population, by using the estimator

$$\bar{x}_L = \bar{x} + B(\bar{Y} - \bar{y}). \tag{3.16}$$

The estimator, $\bar{x}_L$, is called the *linear regression estimator* of $\bar{X}$. Similarly $N\bar{x}_L$ is the linear regression estimator of the population total, X_T.

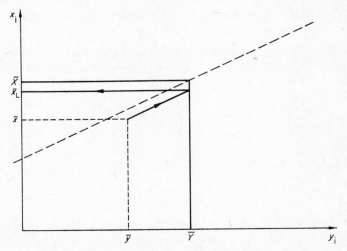

Figure 5   Compensation effect of linear regression estimator $(B > 0; \bar{y} < \bar{Y})$.

Let us consider the justification for this estimator. In the extreme case (3.14) it is incontrovertible, for $\bar{x}_L \equiv \bar{X}$. But no estimation problem exists in this case! For the more general model (3.15), $\bar{x}_L$ again seems a plausible choice. If $B > 0$, then if $\bar{y} < \bar{Y}$ we would expect that $\bar{x} < \bar{X}$ and (3.16) applies a compensation in precisely the manner we would wish, to take account of the linear dependence of X on Y. See Figure 5.

The same is true if $\bar{y} > \bar{Y}$, or if $B < 0$.

C*

Additional justification arises from considering the sampling behaviour of $\bar{x}_L$. It is clearly *consistent* in the finite population sense, since when $n = N$, $\bar{x}_L = \bar{X}$. We see also that

$$E(\bar{x}_L) = E(\bar{x}) + B[\bar{Y} - E(\bar{y})] = \bar{X},$$

so that $\bar{x}_L$ is unbiased.

Furthermore,

$$\mathrm{Var}(\bar{x}_L) = E[(\bar{x}_L - \bar{X})^2] = E\{[(\bar{x} - \bar{X}) - B(\bar{y} - \bar{Y})]^2\}$$

$$= \frac{1-f}{n}(S_X{}^2 - 2BS_{XY} + B^2 S_Y{}^2) \qquad (3.17)$$

$$= \frac{1-f}{n} S_X{}^2(1 - \rho_{XY}^2). \qquad (3.18)$$

Thus $\mathrm{Var}(\bar{x}_L) \leqslant \mathrm{Var}(\bar{x})$, and the efficiency of $\bar{x}_L$ relative to $\bar{x}$ increases with ρ_{XY}^2, that is, with increase in the correlation between the X and Y values.

Thus $\bar{x}_L$ has obvious advantages. It is unbiased for all sizes of sample; it cannot be less efficient than $\bar{x}$. Also we can obtain an *unbiased estimate* of $\mathrm{Var}(\bar{x}_L)$ from the sample, in the form

$$\frac{1-f}{n}(s_X{}^2 - 2Bs_{XY} + B^2 s_Y{}^2),$$

where $s_X{}^2$, s_{XY}, and $s_Y{}^2$ are the familiar unbiased estimators of $S_X{}^2$, S_{XY}, and $S_Y{}^2$;

$$\text{e.g.} \qquad s_{XY} = \frac{1}{n-1} \sum_{i=1}^{n}(x_i - \bar{x})(y_i - \bar{y}).$$

But the results above are not as conclusive as they might appear, for a variety of reasons. In a practical situation, the exact value of the parameter B will of course be unknown. Furthermore, the model (3.15), with the additional assumption of zero correlation between the E_i and Y_i, is unlikely to hold precisely; we cannot assess its propriety without studying the total population of (X, Y) values, in which case we would have no need for sampling or estimation. So how are we to relate these results to the practical situation?

We merely use our study of the behaviour of $\bar{x}_L$ in the special case above to *motivate* the consideration of the *general linear regression estimator*

$$\bar{x}_L = \bar{x} + b(\bar{Y} - \bar{y}) \qquad (3.19)$$

(for some value of b) as a general principle of estimation. With no specific concern for the nature of any relationship between the X_i and Y_i, we might still ask whether an estimator of the form of (*3.19*) has any special value as an estimator of $\bar{X}$. The fact that it clearly does have such value when (X, Y) satisfy (*3.15*) with $S_{YE} = 0$ and $b = B$ justifies such an enquiry.

We must consider two possibilities: either that b is given some pre-assigned value, or that we seek to estimate an appropriate value from the sample.

(a) Preassigned b

The sampling properties of $\bar{x}_L$ defined by (*3.19*) are already known to us from study of the special form (*3.16*)

We see that, *whatever the value of b*,

$$E(\bar{x}_L) = E(\bar{x})+b[\bar{Y}-E(\bar{y})] = \bar{X}$$

so that *the general linear regression estimator is unbiased.*

Also,

$$\mathrm{Var}(\bar{x}_L) = \frac{1-f}{n}(S_X{}^2-2bS_{XY}+b^2S_Y{}^2) \qquad (3.20)$$

with corresponding unbiased sample estimate

$$\frac{1-f}{n}(s_X{}^2-2bs_{XY}+b^2s_Y{}^2).$$

An obvious question arises: if $\bar{x}_L$ is unbiased for all values of b, for what value of b does it have minimum variance? From (*3.20*) this must occur when

$$bS_Y{}^2 - S_{XY} = 0$$

or

$$b = b_0 = S_{XY}/S_Y{}^2 = \rho_{XY}\frac{S_X}{S_Y},$$

in which case $\mathrm{Var}(\bar{x}_L)$ takes the minimum value

$$\mathrm{Min\ Var}(\bar{x}_L) = \frac{1-f}{n}S_X{}^2(1-\rho_{XY}^2). \qquad (3.21)$$

But this is just the same as (*3.18*), so we conclude that

$$\bar{x} - \rho_{XY}\frac{S_X}{S_Y}(\bar{Y} - \bar{y})$$

is the most efficient estimator of $\bar{X}$ of the form (*3.19*), *irrespective of any possible relationship between* X *and* Y *in the population*. If the model (*3.15*) happens to hold, then the optimum estimator is precisely the one, (*3.16*), which we considered for that model (and which had the practical appeal of applying an appropriate form of compensation to $\bar{x}$).

Occasionally we *may* be prepared to assign some specific value to b. Perhaps other studies of a similar nature have been carried out, and we feel fairly confident in transferring earlier knowledge to the present situation. Or again, the measures X and Y may be such that we antici-pate a particular value for the slope of a linear relationship between them. In such situations we can use the sample approximation for (*3.20*) to assess the precision of our estimator, or compare it in efficiency with the s.r. sample mean, $\bar{x}$. We can also construct approximate confidence intervals for $\bar{X}$ in the usual way, on the assumption of the normality of $\bar{x}$ and $\bar{y}$. (Appropriate conditions for this follow from the discussion in Section 2.4.)

Furthermore, by considering the sample analogue of (*3.21*) we can informally assess how well our regression estimator compares in efficiency with the *best possible* estimator of the form (*3.19*). The relative efficiency is

$$(1-\rho_{XY}^2)\left\{1 - 2b\rho_{XY}\left(\frac{S_Y}{S_X}\right) + b^2\left(\frac{S_Y}{S_X}\right)^2\right\}^{-1},$$

which, for large samples, may be reasonably estimated by substituting sample estimates of S_X, S_Y, and ρ_{XY}. We see that the proportional increase in variance due to using a non-optimal value for b is

$$\frac{\text{Var}(\bar{x}_L)}{\text{Min Var}(\bar{x}_L)} - 1 = \frac{\rho_{XY}^2}{1-\rho_{XY}^2}\left(1 - \frac{b}{b_0}\right)^2, \qquad (3.22)$$

where b_0 is the optimal value, $\rho_{XY}S_X/S_Y$.

(*3.22*) has important implications. Non-optimality of choice of the value of b can produce serious inefficiency in the regression estimator, relative to optimal choice. The relative inefficiency will be greatest in populations where X and Y are highly correlated. If the correlation is modest, choice of the value of b is less crucial, but then the potential gain over using $\bar{x}$ is very much less.

Example 3.3

Consider the expression (*3.22*) when $\rho_{XY} = 0.6$, 0.8, 0.9, 0.95 and $|1 - b/b_0| = 0.5, 0.2, 0.1$. The proportional increases in variance in these situations are:

$|1-b/b_0|$ $\diagdown$ ρ_{XY}	0·6	0·8	0·9	0·95
0·5	0·141	0·444	1·066	2·314
0·2	0·023	0·071	0·171	0·370
0·1	0·006	0·018	0·043	0·093

When $\rho_{XY} = 0.95$, modest discrepancies seriously effect the efficiency of the estimator.

(b) Estimated *b*

If we have no basis for preassigning a value to *b*, *as will most often be the case*, we must consider how we might use the derived sample to suggest an appropriate value. The optimal value of *b*, $\rho_{XY} S_X / S_Y$ expressed in terms of population characteristics, suggests that we might try using the corresponding sample expression

$$\tilde{b} = \frac{s_{XY}}{s_Y^2} = \frac{\sum\limits_{i=1}^{n}(x_i - \bar{x})(y_i - \bar{y})}{\sum\limits_{i=1}^{n}(y_i - \bar{y})^2}. \tag{3.23}$$

(This has the same form as the least squares estimator of the classical linear regression coefficient for infinite populations.)

Indeed this is a reasonable procedure, at least in large samples. The regression estimator of the population mean now has the form

$$\bar{x}_L = \bar{x} + \tilde{b}(\bar{Y} - \bar{y}). \tag{3.24}$$

Its distributional properties are difficult to determine precisely in view of the presence of the additional random variable $\tilde{b}$, which itself is a ratio of two statistics.

Large sample behaviour of (*3.24*) is more easily studied, and we base our enquiries on the model (*3.15*) which does not demand any strict linearity of relationship between X_i and Y_i. The extent to which a linear relationship is present will be reflected in the value of ρ_{XY} (as discussed above).

On the basis of this model we can show that the asymptotic forms of the expectation and variance of (*3.24*) are

$$E(\bar{x}_L) = \bar{X} + 0(n^{-1})$$

and

$$\boxed{\text{Var}(\bar{x}_L) = \frac{1-f}{n} S_X{}^2(1 - \rho_{XY}^2) + 0(n^{-\frac{3}{2}})} \qquad (3.25)$$

These results are comforting. Firstly we obtain an estimator which is unbiased in large samples. Secondly, and perhaps more importantly, we see that having to estimate b from the data is no disadvantage *in large samples*. We will obtain the optimum estimator: one with asymptotically as small a variance as is possible for this type of estimator. Thus, using $\hat{b}$ must be preferable to assigning some specific value to b, since at best this will yield an estimator with variance given by the leading term in (*3.25*), whilst if we are unfortunate in our choice of a value for b we may be faced with a far less efficient estimator.

But if all is well in large samples, the same cannot be claimed for small samples. The distinction between 'large' and 'small' samples in this respect requires a more detailed knowledge of how the mean and variance of $\bar{x} - \hat{b}(\bar{Y} - \bar{y})$ depend on sample size and population characteristics. We shall not attempt to derive such results; indeed there is much that is not fully known about the sampling distribution of the regression estimator. We shall merely quote some of the known results. Further details on their derivation and implications can be found in the standard texts and research literature. Again, Cochran (1963, Chapter 7) is a useful source of information and references.

On the question of bias, it happens that this becomes serious if there is marked evidence of a *quadratic* relationship between X and Y. Alternatively, it is aggravated by excess *kurtosis* in the set of Y values in the population. Correspondingly, the variance is seen to be most affected by the coefficient of *skewness* of the Y values, so that the large sample approximation will be least accurate for highly skew populations (with respect to the Y variable). A reasonable practical prescription for adopting the large sample approximation for the variance of the regression estimator is that the sample size should be in excess of 50, and the set of Y values in the population not greatly skew.

Again, in use we will need to *estimate* the variance

$$\frac{1-f}{n} S_X{}^2(1 - \rho_{XY}^2),$$

and we will use

$$s^2(\bar{x}_L) = \frac{1-f}{n}(s_X{}^2 - \tilde{b}s_{XY})$$

or possibly

$$\left(\frac{n-1}{n-2}\right)s^2(\bar{x}_L)$$

by analogy with the unbiased residual variance estimate in classical linear regression estimation for infinite populations.

3.4 Comparison of Ratio and Regression Estimators

We have considered the circumstances under which the *ratio estimator*, $\bar{x}_R$, of $\bar{X}$ is more efficient than the s.r. *sample mean*, $\bar{x}$. This arises if

$$\rho_{XY} > \frac{R\,S_Y}{2\,S_X} \text{ (where } R = \bar{X}/\bar{Y}). \tag{3.26}$$

We saw that, for large samples,

$$\text{Var}(\bar{x}_R) \doteqdot \frac{1-f}{n}(S_X{}^2 - 2R\,\rho_{XY}\,S_X S_Y + R^2 S_Y{}^2). \tag{3.27}$$

Correspondingly, we observed that the *regression estimator* $\bar{x}_L$ (with b estimated from the data) has

$$\text{Var}(\bar{x}_L) \doteqdot \frac{1-f}{n}S_X{}^2(1-\rho_{XY}^2), \tag{3.28}$$

and that it *cannot be less efficient than* $\bar{x}$.

There remains the comparison of $\bar{x}_R$ and $\bar{x}_L$. We have

$$\text{Var}(\bar{x}_R) - \text{Var}(\bar{x}_L) \doteqdot \frac{1-f}{n}(R^2 S_Y{}^2 - 2R\rho_{XY}\,S_X S_Y + \rho_{XY}{}^2\,S_X{}^2)$$

$$= \frac{1-f}{n}(RS_Y - \rho_{XY}\,S_X)^2, \tag{3.29}$$

and we therefore conclude that *the regression estimator must be at least as efficient as the ratio estimator* under all circumstances. From (3.29) we see that the only situation in which the ratio estimator can have the same efficiency as the regression estimator is when

$$R = \rho_{XY}\frac{S_X}{S_Y}. \tag{3.30}$$

But we have seen that $\rho_{XY}S_X/S_Y$ is the optimum choice b_0 for the parameter b, in the sense that it minimises the variance of $\bar{x}+b(\bar{Y}-\bar{y})$. We saw also that it is the inevitable value for the parameter B in the model (*3.15*). Thus $\bar{x}_R$ and $\bar{x}_L$ are equally efficient only if

$$R = b_0 = B.$$

Let us look a little more closely at the comparison of $\bar{x}$, $\bar{x}_R$, and $\bar{x}_L$, and at the role of any formal model expressing an element of linearity (or proportionality) in the relationship between the X and Y values in the population.

Note immediately that *we do not need to make any explicit assumptions about a possible linear relationship between X and Y to derive the properties of $\bar{x}$, $\bar{x}_R$, and $\bar{x}_L$ described above*. Merely defining $\bar{x}_R$ by (*3.8*) and $\bar{x}_L$ by (*3.24*), we find that (asymptotically, hence approximately in large samples) they are unbiased and have variances given by (*3.27*) and (*3.28*), respectively. Thus $\bar{x}_L$ is always more efficient than $\bar{x}$, except in the isolated case where $\rho_{XY} = 0$, when they are equally efficient. Also, the relative efficiencies of $\bar{x}_R$ and $\bar{x}$ are governed by the value of ρ_{XY}: $\bar{x}_R$ being more efficient if (*3.26*) is satisfied (with the prerequisite that $C_Y \leqslant 2C_X$), otherwise less efficient. Finally, $\bar{x}_L$ will always be more efficient than $\bar{x}_R$, unless the special relationship (*3.30*) happens to hold in the population, in which case they are equally efficient. If (*3.30*) holds, then so of course does (*3.26*) and the conditions are satisfied for $\bar{x}_R$ to be more efficient than $\bar{x}$. (Otherwise we could encounter a contradiction where $\bar{x}_L$ and $\bar{x}_R$ had the same variance, but $\bar{x}_L$ was inevitably more efficient than $\bar{x}$, whilst $\bar{x}_R$ happened to be less efficient than $\bar{x}$!)

Thus, what are important in this comparison are the *relative* values of $\bar{X}$ and $\bar{Y}$, and of $S_X{}^2$ and $S_Y{}^2$, and the correlation coefficient, ρ_{XY}. We have no need to formulate any linear model to express the relationship between X and Y. However, we could choose to set up such a model, and this would serve two purposes. Firstly, it would provide a practical motivation for initially considering types of estimator of the form of $\bar{x}_R$ or $\bar{x}_L$, as we have already seen. Secondly, since ρ_{XY} is a measure of *linear* association, such a model might help to illustrate more tangibly the comparison of $\bar{x}_R$ and $\bar{x}_L$.

With no implied constraints on the bivariate finite population of values (X_i, Y_i) we can freely declare that

$$X_i = \bar{X}+k(Y_i-\bar{Y})+E_i, \tag{3.31}$$

or that

$$X_i = k'Y_i+E_i', \tag{3.32}$$

with

$$\sum_{i=1}^{N} E_i = \sum_{i=1}^{N} E_i' = 0.$$

It must of course be true that $k' = R$.

But with no additional assumptions, such models are sterile; they imply no linearity or proportionality of relationship of X and Y. However complex the pattern of values (X_i, Y_i), this can be accommodated by suitable values E_i (or E_i') depending on the Y_i. But if we demand that such dependence is not to be entertained, say by postulating that

$$\sum_{i=1}^{N} E_i(Y_i - \bar{Y}) = \sum_{i=1}^{N} E_i'(Y_i - \bar{Y}) = 0,$$

the models become much more structured. They now represent linearity, or proportionality, with superimposed 'deviations' (or 'errors') uncorrelated with the Y values.

As we have seen at the beginning of the previous section, we must now have, in (3.31),

$$k = B = \rho_{XY} \frac{S_X}{S_Y}.$$

Also in (3.32) we find, on multiplying each side by $(Y_i - \bar{Y})$ and summing over the whole population,

$$k' = R = \rho_{XY} \frac{S_X}{S_Y}.$$

Thus if (3.32), with

$$\sum_{i=1}^{N} E_i'(Y_i - \bar{Y}) = 0,$$

happens to be an appropriate model for our population, it can be re-expressed as (3.31) with $E_i = E_i'$, and we have

$$R = k' = k = B.$$

But this is precisely the condition that needs to hold for $\bar{x}_R$ and $\bar{x}_L$ to be equally efficient. Hence from the practical viewpoint this means that the tangible justification for using $\bar{x}_R$ will rest on any indication in the data of a *linear relationship through the origin* between the values of X and Y, with no suggestion of correlation between the Y_i and the deviations, E_i. So such a relationship does turn out to have a formal basis. The likely inferiority of $\bar{x}_R$ relative to $\bar{x}_L$ will be indicated by observing that the X and Y values do not seem to be roughly *proportional* (in a positive sense) to one another—or do not appear even to be roughly linearly related.

In conclusion, we can informally summarise the conditions which will support the use of ratio or regression estimators in the following way.

We are concerned with estimating $\bar{X}$ (or $\bar{X}_T$) in a finite population where for each X_i value there is a value Y_i of some auxiliary variable. X and Y can be simultaneously sampled and the population mean $\bar{Y}$ is known precisely. If the data (or general knowledge) suggest some reasonable degree of linear relationship between the X and Y values then we can expect to obtain a useful gain in efficiency over $\bar{x}$ (or x_T) by using the regression estimator $\bar{x}_L$ (or $x_{TL} = N\bar{x}_L$). If the linear relationship has positive slope and appears to pass through the origin, we can expect a similar gain, for less computational effort, by using the ratio estimator $\bar{x}_R$ (or x_{TR}). This saving in computation can be the only advantage of $\bar{x}_R$ over $\bar{x}_L$, and can only be worth exploiting if there *is* an indication of a positive linear relationship through the origin.

Note that in our discussion and comparison of these estimators the sole concern has been for achieving (asymptotically) unbiasedness and minimum variance. Whilst we shall not consider here any other criteria of choice, we should not disregard alternative prospects. Questions of the cost involved in achieving a certain precision are most relevant. In our study (in the next chapter) of estimators for stratified populations, we shall have cause to consider cost optimality as a criterion of choice in addition to the idea of minimising the variance of unbiased estimators.

Exercises for Chapter 3

I

Part of a coniferous forest contains 280 trees of the same species and of similar ages. A preliminary estimate is required of the total weight of timber that these trees will yield. A forestry expert claims to be able to make fairly accurate assessments of the yield from any tree merely by visual inspection, and makes such assessments for all 280 trees. He assessed the total yield at 432·6 tons. Subsequently, 25 trees picked at random are felled and their timber yields accurately determined. The actual yields, x_i, and corresponding assessed yields, y_i, provide the following summary results.

$$\sum_1^{25} x_i = 39\cdot2, \quad \sum_1^{25} y_i = 40\cdot7,$$

$$\sum_1^{25} x_i^2 = 66\cdot92, \quad \sum_1^{25} x_i y_i = 68\cdot43, \quad \sum_1^{25} y_i^2 = 71\cdot17$$

Estimate the total yield using either the ratio or regression estimator, whichever seems most appropriate. Compare the efficiencies of the ratio estimator, the regression estimator, and the estimator based on the sample of x_i values alone.

II

In studying lung function in a group of 560 workers in a coal mine, an estimate was required of the mean value of some relevant measure X. A s.r. sample of 10 workers was chosen and their X values, x_i, determined by an appropriate test. A note was also made of their heights, y_i. The results were:

x_i	3·0	3·5	3·3	3·1	4·1	3·2	3·7	2·9	3·9	3·4
y_i (in)	68	72	67	69	63	62	66	71	70	64

From routine medical records the average height for the group of 560 workers is known to be $\bar{Y} = 68·2$ in. Estimate $\bar{X}$ from the data, and calculate an approximate standard error for your estimator.

III

A sum of £4000 is available to conduct a survey of weekly expenditure on food by households in a particular town. A list of households is available; there are 8220 in all. The enquiry is to relate to some specific week of the year. Interview cost to determine the week's expenditure (x_i) for a randomly chosen household is £3·25; it costs an additional £0·25 to obtain the value (y_i) of some other relevant concomitant variable. The cost of setting up the survey is £500.

The population mean $\bar{Y}$ is known to be 2·9. Previous experience suggests that other relevant population characteristics will have *approximately* the values quoted:

$$C_X = 0·3, \quad C_Y = 0·2, \quad \rho_{XY} = 0·45, \quad R = 3·1.$$

Determine whether it would be better to use the ratio estimator or s.r. sample mean to estimate $\bar{X}$, if processing and calculation costs are £100 and £50, respectively. What size of sample should be used?

4 Stratified Populations

In our earlier enquiries about the accuracy with which we can estimate a population characteristic, $\bar{X}$ say, from the analogous quantity $\bar{x}$ in a s.r. sample, we saw that a crucial factor is the variance S_X^2 of the population. The larger the dispersion in the population, reflected in the value of S_X^2, the less accurate the estimator $\bar{x}$ in the sense that its sampling variance is larger. This is only to be expected; it makes sense intuitively, it is a feature of classical statistical method for infinite populations, it is borne out in the sampling variance, $(1-f)S_X^2/n$, of $\bar{x}$ in finite populations.

Consider a simple numerical example. Suppose we have a finite population of 20 members in which X takes values

$$6 \quad 3 \quad 4 \quad 4 \quad 5 \quad 3 \quad 6 \quad 2 \quad 3 \quad 2 \quad 2 \quad 6 \quad 5 \quad 3 \quad 5 \quad 2 \quad 4 \quad 6 \quad 4 \quad 5$$

Its mean is $\bar{X} = 4$; its variance, $S_X^2 = 40/19$. If we take a s.r. sample of size 5 and use the s.r. sample mean $\bar{x}$ to estimate $\bar{X}$ we have

$$\mathrm{Var}(\bar{x}) = 6/19.$$

Clearly we could obtain quite a range of different values for $\bar{x}$ in different samples; from 2.2 to 5.8. But notice the *structure* of the population. It could be rearranged as

$$2 \quad 2 \quad 2 \quad 2 \quad 3 \quad 3 \quad 3 \quad 3 \quad 4 \quad 4 \quad 4 \quad 4 \quad 5 \quad 5 \quad 5 \quad 5 \quad 6 \quad 6 \quad 6 \quad 6$$

It consists of 5 groups, in each of which all 4 X-values are the same. Suppose we had some mechanism by which we could choose one member at random from *each* group to constitute our sample of size 5. We must inevitably obtain *on all occasions*

$$2, \ 3, \ 4, \ 5, \ 6,$$

with sample mean 4. Thus our estimate has *no sampling fluctuation*, and is always equal to the population mean $\bar{X}$ it is estimating. This extremely favourable situation has arisen because we have been able to *remove all variability* from within the groups in which we are drawing our random observations.

Consider a more practical example. We want to conduct a survey to estimate the mean height of the schoolchildren in a small primary school with four classes, each of about 30 children, covering four different age groups. We decide to measure the heights of a sample of 20 children for this purpose. We might do that by attempting to pick the children at random from the playground during their brief mid-morning break. But there would not be much time for this, and the natural age (or class) groupings of children at play would raise certain difficulties for the choice of a *random* sample. For sheer convenience (and with a useful by-product of providing natural teaching material) it might be better to visit the four classes at lesson time and measure the heights of a s.r. sample of 5 children from each class. The prime stimulus for this approach is convenience or ease of sampling, but note the effect. The classes reflect natural groupings of the population and, because of the relationship between stature and age, the heights in each such group are likely to be less variable than in the population at large. This effect will be by no means as extreme as in the numerical example, but we might well wonder whether the relative homogeneity of the groups might not lead to *some* improvement in the efficiency of our estimate of the mean height, compared with a s.r. sample drawn from the total population.

These examples suggest two possible factors relating to the study of grouped, or *stratified*, populations. Firstly, that stratification might be an aid to efficient estimation under appropriate conditions; secondly, that stratification may be inevitable or expedient in administrative terms.

It would appear that we may be able to estimate some population characteristic more efficiently by sampling each group (*stratum*) separately than by sampling the population at large, if it so happens that our variable X shows less variation within each stratum than in the total population. The opposite effect also seems intuitively plausible. There is no reason why any administratively induced stratification should *necessarily* yield the desirable within-stratum homogeneity. There must also be some limit (due to administrative constraints and to our inevitable lack of knowledge of the detailed form of the population) to the extent to which we can engender such homogeneity by a deliberately chosen stratification of the population.

Clearly we should consider the situation in more detail. In this chapter we will study how to estimate population characteristics in stratified populations, under what circumstances we can expect to obtain better estimators than those derived from a s.r. sample from the unstratified population, and the extent to which practical considerations may influence any potential efficiency gains from stratification.

4.1 Stratified Random Sampling

Suppose we wish to estimate the mean, $\bar{X}$, of the set of values X_1, $X_2, \ldots, X_N$ in a finite population. We shall assume that the population is *stratified*, that is to say it has been divided into k non-overlapping groups, or *strata*, of sizes

$$N_1, N_2, \ldots, N_k \ \left(\sum_{i=1}^{k} N_i = N\right)$$

with members

$$X_{ij} \ (i = 1, \ldots, k; \ \ j = 1, \ldots, N_i).$$

Thinking of each stratum as a sub-population we can carry over our earlier notation, and denote the stratum means and variances by

$$\bar{X}_1, \bar{X}_2, \ldots, \bar{X}_k,$$

and

$$S_1{}^2, S_2{}^2, \ldots, S_k{}^2,$$

respectively.

The population mean and variance, $\bar{X}$ and S^2, will of course have the forms

$$\bar{X} = \frac{1}{N} \sum_{i=1}^{k} N_i \bar{X}_i = \sum_{i=1}^{k} W_i \bar{X}_i$$

(where W_i is termed the *weight* of the ith stratum), and

$$S^2 = \frac{1}{N-1} \left\{ \sum_{i=1}^{k} \sum_{j=1}^{N_i} (X_{ij} - \bar{X}_i + \bar{X}_i - \bar{X})^2 \right\}$$

$$= \frac{1}{N-1} \left\{ \sum_{i=1}^{k} (N_i - 1) S_i{}^2 + \sum_{i=1}^{k} N_i (\bar{X}_i - \bar{X})^2 \right\}. \quad (4.1)$$

We shall assume that a sample of size n is chosen *by taking a s.r. sample of pre-determined size from each stratum*. The stratum sample sizes will be denoted $n_1, n_2, \ldots, n_k$ ($\Sigma_{i=1}^{k} n_i = n$). The s.r. sample from the ith stratum has members

$$x_{i1}, x_{i2}, \ldots, x_{in_i} \ (i = 1, 2, \ldots, k),$$

and we denote the sample mean and variance in the ith stratum by

$$\bar{x}_i = \frac{1}{n_i} \sum_{j=1}^{n_i} x_{ij}$$

and

$$s_i^2 = \frac{1}{n_i - 1} \sum_{j=1}^{n_i} (x_{ij} - \bar{x}_i)^2.$$

In each stratum we have a *sampling fraction*, $f_i = (n_i/N_i)(i = 1, 2, ..., k)$.

Such a sampling procedure for the choice of a sample of total size n from the overall population is termed *stratified* (simple) *random sampling*.

At this stage we shall make no assumptions or enquiries about the basis of the stratification of the population—we shall just accept that the population is so stratified, and that we wish to estimate $\bar{X}$ from the sample values x_{ij} yielded by stratified random sampling. Later, we shall consider how the stratification might be constrained by administrative considerations (sampling ease, costs, etc.) or, in contrast, how we can sometimes make use of any practical information we may possess about the population to effect a stratification likely to lead to particularly efficient estimation of $\bar{X}$.

The estimator of $\bar{X}$ commonly employed is the so-called *stratified sample mean*. This is defined as

$$\bar{x}_{st} = \sum_{i=1}^{k} W_i \bar{x}_i.$$

Note: this assumes that we know the stratum sizes N_i precisely, in order to determine the stratum weights $W_i = N_i/N$.

The stratified sample mean, $\bar{x}_{st}$, is not the same as the arithmetic mean

$$\bar{x}' = \frac{1}{n} \sum_{i=1}^{k} n_i \bar{x}_i$$

of the stratified random sample, except under the special circumstances where

$$\frac{n_i}{n} = \frac{N_i}{N}.$$

This would imply that *the sampling fractions* $f_i = n_i/N_i$ *are identical for all strata*; a special form of stratified sampling where the stratum sample sizes, n_i, are said to be chosen by *proportional allocation* (since the sample sizes are chosen to be *proportional* to the stratum sizes). Such a principle can be time-saving with regard to the collection of the sample data, and we shall see that it can have statistical advantages, but again it presupposes a knowledge of the stratum sizes, N_i. We assume throughout that this knowledge exists; otherwise the weights W_i will need to be *estimated* in some manner, with a resulting bias in the estimator $\bar{x}_{st}$ and loss of accuracy in the later derived results. Further comments on this problem are made in Section 4.5.

We must firstly consider the mean and variance of $\bar{x}_{st}$. We have

$$E(\bar{x}_{st}) = \sum_{i=1}^{k} W_i \bar{X}_i = \bar{X}, \text{ so that } \bar{x}_{st} \text{ is } unbiased,$$

in view of the inevitable unbiasedness of the stratum sample means, $\bar{x}_i$. Note that

$$E(\bar{x}') = \frac{1}{n} \sum_{i=1}^{k} n_i \bar{X}_i,$$

so that $\bar{x}'$ *will be unbiased only in the case of proportional allocation* (where $n_i/n = N_i/N$).

For the variance of $\bar{x}_{st}$ we have

$$\mathrm{Var}(\bar{x}_{st}) = \sum_{i=1}^{k} W_i^2 (1-f_i) S_i^2 / n_i \qquad (4.2)$$

The proof is straightforward:

$$\mathrm{Var}(\bar{x}_{st}) = \sum_{i=1}^{k} W_i^2 \, \mathrm{Var}(\bar{x}_i)$$

provided (as is implied in the stratified random sampling procedure) that $\mathrm{Cov}(\bar{x}_i, \bar{x}_j) = 0$ if $i \neq j$, that is, the simple random sample means for different strata are uncorrelated. Thus we obtain (*4.2*), using (*2.3*) for $\mathrm{Var}(\bar{x}_i)$.

One further aspect of the behaviour of $\bar{x}'$ needs comment. Although with proportional allocation $\bar{x}_{st}$ and $\bar{x}'$ coincide, the arithmetic mean $\bar{x}'$ does *not* have the same variance as $\bar{x}$, the mean of a s.r. sample of size n from the whole population. The variance of $\bar{x}'$ takes on the appropriate special form of (*4.2*) (see (*4.4*) below), whilst

$$\mathrm{Var}(\bar{x}) = (1-f) S^2 / n,$$

with $f = n/N$. The reason for this discrepancy is the element of non-randomness of the stratified random sample that arises from the constraint that specific numbers, n_i, of members of the sample *must* be chosen from each distinct sub-population defined by the stratification.

Some special cases of (4.2) should be considered.

(a) *Sampling fractions* $f_i = n_i/N_i$ *negligible,*

$$\text{Var}(\bar{x}_{st}) = \sum_{i=1}^{k} W_i^2 S_i^2/n_i. \qquad (4.3)$$

(b) *Proportional allocation,* $n_i = nW_i, f_i = f = n/N,$

$$\text{Var}(\bar{x}_{st}) = \frac{(1-f)}{n} \sum_{i=1}^{k} W_i S_i^2. \qquad (4.4)$$

(c) *Proportional allocation, and constant within strata variances,*
$S_i^2 = S_W^2 (i = 1, 2, ..., k)$

$$\text{Var}(\bar{x}_{st}) = \frac{(1-f)}{n} S_W^2. \qquad (4.5)$$

Directly analogous results hold for the estimation of the population total, X_T. Thus, from the stratified random sample we obtain an unbiased estimator $N\bar{x}_{st} = \Sigma_{i=1}^{k} N_i\bar{x}_i$, with variance $\Sigma_{i=1}^{k} N_i^2(1-f_i)S_i^2/n_i$.

In practical situations the stratum variances, S_i^2, will not be known. So if we wish to quote a standard error for the estimator $\bar{x}_{st}$, or to construct an approximate confidence interval for $\bar{X}$, based on $\bar{x}_{st}$ (and involving its approximate normality in large samples), we will *need to estimate the* S_i^2. We have already considered this problem in the discussion (Section 2.3) of how to obtain an unbiased estimator of a population variance from a s.r. sample. The strata are just sub-populations, the sampled values in each stratum constitute a s.r. sample.

Thus, as

$$s_i^2 = \frac{1}{n_i-1} \sum_{j=1}^{n_i} (x_{ij}-\bar{x}_i)^2 \qquad (i = 1, 2, ..., k)$$

are (*unbiased*) estimators of the stratum variances $S_i^2(i = 1, 2, ..., k)$, we obtain *an unbiased estimator of* $\text{Var}(\bar{x}_{st})$ as

$$s^2(\bar{x}_{st}) = \sum_{i=1}^{k} W_i^2(1-f_i)s_i^2/n_i$$

$$= \frac{1}{N^2} \sum_{i=1}^{k} N_i(N_i-n_i)s_i^2/n_i. \qquad (4.6)$$

Naturally this requires that *all stratum sample sizes should be at least 2*, i.e. $n_i \geq 2$ $(i = 1, 2, \ldots, k)$.

It is not entirely unreasonable that occasionally we may encounter some $n_i = 1$, typically when the population is highly variable and a very large number of strata need to be considered. Estimation of $\text{Var}(\bar{x}_{st})$ is *not* hopeless in such situations: two ingenious methods of dealing with the extreme case where all $n_i = 1$ are described briefly by Cochran (1963, Section 5A.11).

In some situations the practical circumstances may suggest that *all* stratum variances are equal. If this is so, then it is desirable to combine the data from the different strata to obtain an overall, or 'pooled', unbiased estimator of the common variance S_W^2 in the form

$$s_W^2 = \frac{1}{N-k} \sum_{i=1}^{k} \sum_{j=1}^{n_i} (x_{ij} - \bar{x}_i)^2.$$

We can now estimate $\text{Var}(\bar{x}_{st})$ by

$$s^2(\bar{x}_{st}) = \frac{s_W^2}{N^2} \sum_{i=1}^{} N_i(N_i - n_i)/n_i.$$

In such a situation we will frequently strive (for reasons made clear later) to use proportional allocation in drawing the sample, and we then have simply

$$s^2(\bar{x}_{st}) = \left(1 - \frac{n}{N}\right) s_W^2/n$$

as an unbiased estimator of $\text{Var}(\bar{x}_{st})$.

Approximate confidence intervals for $\bar{X}$ (or X_T) may again be constructed on the assumption of a normal sampling distribution for the estimator, $\bar{x}_{st}$. These will be given (for confidence level $1 - \alpha$) by

$$\bar{x}_{st} - z_\alpha s(\bar{x}_{st}) < \bar{X} < \bar{x}_{st} + z_\alpha s(\bar{x}_{st})$$

or

$$N[\bar{x}_{st} - z_\alpha s(\bar{x}_{st})] < X_T < N[\bar{x}_{st} + z_\alpha s(\bar{x}_{st})].$$

As before, such approximations will only be reasonable if the conditions are satisfied (in terms of sample size, and so on) for $\bar{x}_{st}$ to have a distribution which is essentially normal, and for $s^2(\bar{x}_{st})$ to be close in value to $\text{Var}(\bar{x}_{st})$. The latter requirement is often the more stringent one, and replacement of z_α by a percentage point for an appropriate t-distribution can lead to greater accuracy. But in stratified random sampling, the situation is far less clear-cut than in s.r. sampling from the total population, and no general prescription for the construction of approximate confidence intervals is available. One difficulty is that although *total* sample size may be large, the s.r. samples within each stratum will frequently not be. Some work has to be done on what constitutes an 'appropriate' number of degrees of freedom to adopt when using a *t*-distribution in place of the normal distribution, but it can hardly be claimed to lead to any universally applicable policy.

4.2 Comparison of the Simple Random Sample Mean and the Stratified Sample Mean

In the introduction to this chapter it was suggested that stratification of the population may on occasions increase the efficiency with which we can estimate population characteristics such as $\bar{X}$ or X_T. To examine this possibility let us compare $\bar{x}$ and $\bar{x}_{st}$ in the same situation. Both are unbiased estimators. Which is the more efficient, in the sense of having the smaller variance? We know that

$$\text{Var}(\bar{x}) = (1-f)S^2/n,$$

whilst $\text{Var}(\bar{x}_{st})$ is given by (*4.2*). To simplify the comparison we shall suppose that the stratified sample has been drawn *with proportional allocation*. Then, from (*4.4*),

$$\text{Var}(\bar{x}_{st}) = \frac{1-f}{n} \sum_{i=1}^{k} \frac{N_i}{N} S_i^2$$

and

$$\text{Var}(\bar{x}) - \text{Var}(\bar{x}_{st}) = \frac{(1-f)}{n} \left(S^2 - \frac{1}{N} \sum_{i=1}^{k} N_i S_i^2 \right). \qquad (4.7)$$

But, by (*4.1*),

$$S^2 = \frac{1}{N-1} \left\{ \sum_{i=1}^{k} (N_i-1)S_i^2 + \sum_{i=1}^{k} N_i(\bar{X}_i - \bar{X})^2 \right\}.$$

Now if the stratum sizes, N_i, are large enough

$$\frac{N_i-1}{N-1} \doteqdot \frac{N_i}{N} \doteqdot \frac{N_i}{N-1} \qquad (4.8)$$

and

$$S^2 \doteqdot \frac{1}{N} \left\{ \sum_{i=1}^{k} N_i S_i^2 + \sum_{i=1}^{k} N_i(\bar{X}_i - \bar{X})^2 \right\},$$

so that from (*4.7*)

$$\text{Var}(\bar{x}) - \text{Var}(\bar{x}_{st}) = \frac{(1-f)}{nN} \sum_{i=1}^{k} N_i(\bar{X}_i - \bar{X})^2$$

$$= \frac{(1-f)}{n} \sum_{i=1}^{k} W_i(\bar{X}_i - \bar{X})^2, \qquad (4.9)$$

which is positive, unless, exceptionally, the $\bar{X}_i$ are all the same. Thus it appears that the stratified sample mean will always be more efficient than the s.r. sample mean, the more so the larger the variation in the stratum means.

But the assumption (*4.8*) proves to be quite crucial. Suppose the stratum sample sizes are not large enough for (*4.8*) to hold. Then, using (*4.1*) for S^2, we find

$$\text{Var}(\bar{x}) - \text{Var}(\bar{x}_{st}) = \frac{(1-f)}{n(N-1)} \left\{ \sum_{i=1}^{k} N_i(\bar{X}_i - \bar{X})^2 \right.$$
$$\left. - \frac{1}{N} \sum_{i=1}^{k} (N-N_i)S_i^2 \right\}, \qquad (4.10)$$

which need not necessarily be positive. This more precise comparison shows that $\bar{x}_{st}$ is *not* necessarily more efficient than $\bar{x}$ under all circumstances; $\bar{x}_{st}$ *will be more efficient than* $\bar{x}$ *if*

$$\sum_{i=1}^{k} N_i(\bar{X}_i - \bar{X})^2 > \frac{1}{N} \sum_{i=1}^{k} (N-N_i)S_i^2.$$

A further specialisation gives a more tangible expression of this condition. Suppose all the strata have the same variance, S_W^2. Then we require

$$\frac{1}{k-1} \sum_{i=1}^{k} N_i(\bar{X}_i - \bar{X})^2 > S_W^2. \qquad (4.11)$$

Thus, the stratified sample mean will be *more efficient* than the s.r. sample mean *if variation between the stratum means is sufficiently large compared with within-strata variation*; the greater this advantage the greater the efficiency of $\bar{x}_{st}$ relative to $\bar{x}$. (Note how this comparison mirrors the analysis of variance criterion in infinite normal samples for testing homogeneity of a set of means.)

Summarising the results of this section we can informally conclude that the higher the variability in stratum means, and the lower the variability of X values within the strata, the greater is the potential gain from using the stratified sample mean $\bar{x}_{st}$ (rather than $\bar{x}$) for estimating $\bar{X}$. The same will be true, of course, for the estimation of X_T.

Example 4.1

It is interesting to see if we can illustrate the properties of $\bar{x}_{st}$, which are discussed above, by stratifying the population of heights given in the *Class Example* and studying corresponding stratified sample means. There are certain obvious methods of stratification in this situation, namely

   (i) by rows,
   (ii) by columns,
   (iii) by sex.

Thus in (i) and (ii) we have 5 strata of equal size, each stratum consisting of the 5 students sitting in some particular row, or column, respectively. (See Figure 1, where the row-strata are labelled I, ... V; the column strata A, B, C, D, E.) In (iii) we have two strata, male and female students, with 15 and 10 members respectively. Suppose we take a stratified random sample of size 5, with proportional allocation, in each case. That is, in

   (i) we pick one student at random from each row,
   (ii) we pick one student at random from each column,
   (iii) we take a s.r. sample of 3 male, and a s.r. sample of 2 female, students.

Comparison of the sampling distributions of $\bar{x}_{st}$ (in each case) with earlier derived empirical sampling distributions for $\bar{x}$, and for $\bar{x}_R$, should prove interesting. Relevant population characteristics are

(i) $\bar{X}_I = 3\cdot4$ $\bar{X}_{II} = 7\cdot2$ $\bar{X}_{III} = 9\cdot2$
     $\bar{X}_{IV} = 11\cdot8$ $\bar{X}_V = 10\cdot6$

     $S_I^2 = 3\cdot3$ $S_{II}^2 = 1\cdot2$ $S_{III}^2 = 2\cdot2$
     $S_{IV}^2 = 6\cdot2$ $S_V^2 = 7\cdot3$

(ii) $\bar{X}_A = 9\cdot0$ $\bar{X}_B = 8\cdot0$ $\bar{X}_C = 7\cdot2$
     $\bar{X}_D = 8\cdot4$ $\bar{X}_E = 9\cdot6$

     $S_A^2 = 10\cdot5$ $S_B^2 = 14\cdot0$ $S_C^2 = 6\cdot7$
     $S_D^2 = 25\cdot3$ $S_E^2 = 13\cdot8$

(iii) $\bar{X}_M = 10\cdot4$ $\bar{X}_F = 5\cdot5$
     $S_M^2 = 6\cdot54$ $S_F^2 = 6\cdot94$

Five hundred stratified random samples of size 5 have been chosen in each case, and figures 6, 7 and 8 present the three histograms of values of $\bar{x}_{st}$ so obtained. The approximate variances of $\bar{x}_{st}$, estimated from the 500 values of x_{st} in each case, are

(i)   0·676   (0·646)
(ii) 2·014   (2·250)
(iii) 1·064   (1·073)

The values in brackets are those obtained directly from the theoretical form (4.4) using the appropriate population characteristics.

We see that stratification by rows produces a dramatic increase in efficiency over the s.r. sample mean, $\bar{x}$—whose variance is 1·99. The efficiency of $\bar{x}_{st}$, is at least as good as (possibly slightly higher than) that of the ratio estimator, $\bar{x}_R$, which has estimated variance 0·685. Likewise, stratification by sex provides an improvement over $\bar{x}$, but is not superior to the use of $\bar{x}_R$. Stratification by columns is clearly no advantage; indeed we will do better even to use the s.r. sample mean, $\bar{x}$.

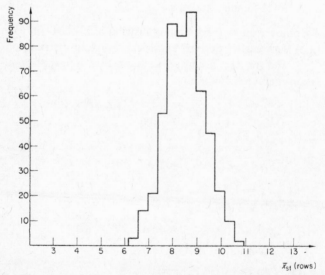

Figure 6   Histogram of 500 values of x_{st} (rows) for samples of size 5 in the *Class Example*.

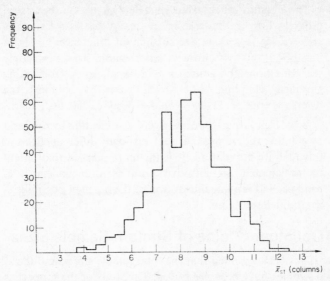

Figure 7   Histogram of 500 values of $\bar{x}_{st}$ (columns) for samples of size 5 in the *Class Example*

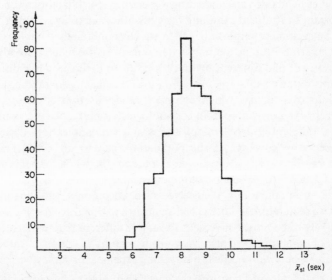

Figure 8   Histogram of 500 values of $\bar{x}_{st}$ (sex) for samples of size 5 in the *Class Example*

These results reflect what we would expect on theoretical grounds. For stratification by rows (and to a lesser extent by sex) we see that there are substantial differences in strata means, but relatively little variation within strata—precisely the conditions for a useful gain in efficiency from using stratified sampling. For stratification by columns the reverse is true, so (as we would expect) $\bar{x}_{st}$ has little merit.

What is puzzling is precisely why stratification by row, and by column, should produce such distinctly different effects in this real-life population. It is far more marked than might be anticipated on intuitive grounds; nonetheless it is genuine, and is a fine illustration of the range of possibilities in stratified sampling.

4.3 Optimum Choice of Stratum Sample Sizes

So far it has been assumed that the total sample size n, and the stratum sample sizes n_i, have been *prescribed*. The study of the properties of $\bar{x}_{st}$ has assumed some specific allocation $n_1, n_2, ..., n_k$. As in all survey sampling schemes, we must remain aware of the need to achieve some required precision of estimation, and to do so either for minimum cost or (if possible) within some cost limitation imposed by the resources available for conducting the enquiry. Such factors are no less important in stratified sampling than for other sampling schemes. They may be even more important. Often we have only limited choice of the basis of stratification, the major determinant is frequently an administrative one: that different methods need to be used to sample different sections of the population, or that natural (financial, geographic, social) divisions exist across which complete randomisation is clumsy, and unnecessary. For example, in a sociological study, lack of a common listing, differential problems of access and communication, desire for representative coverage of the population, and so on, could make it undesirable to attempt to sample at random the whole population (of hospital patients, prisoners, old-age pensioners, etc.). Intrinsic interest in the sub-populations themselves also encourages stratification—a point we shall return to later. Then again, in a national geographic survey, it is likely to be most convenient to sample different regions separately; costs of sampling will also vary from region to region (stratum to stratum) if only with respect to travelling expenses for survey workers.

So, again, we must consider the question of how to choose the sample size n to satisfy certain precision or cost requirements. Since different strata are likely to exhibit different degrees of variability, we must inevitably proceed beyond the choice of n to the allocation of the individual stratum sample sizes, n_i. Also, to merely state our requirements in terms of the *variance* of some estimator will not be sufficient in general. Different sampling costs for different strata imply that we must attempt to take some account of cost factors in determining a desirable allocation of stratum sample sizes.

In any particular problem, local circumstances should enable a reasonably precise statement to be made of sampling costs in the different strata. Simple cost models have been proposed within which a large number of practical problems can be accommodated. The simplest form assumes that there is some overhead cost, c_0, of administering the survey, and that individual observations from the ith stratum each cost an amount c_i. Thus the total cost is

$$C = c_0 + \sum_{i=1}^{k} c_i n_i. \qquad (4.12)$$

This is the model which we shall adopt, although constant unit cost for observations sometimes exaggerates the true situation, and the replacement of $\sum_{i=1}^{k} (c_i n_i)$ by, say, $\sum_{i=1}^{k} (d_i \sqrt{n_i})$ may be better. This latter form is more reasonable, for example, when the major cost ingredient is for travel.

Suppose we adopt the cost model (*4.12*) and ask what allocation of stratum sample sizes, $n_1, n_2, \ldots, n_k$, should be adopted to

(I) minimise $\mathrm{Var}(\bar{x}_{st})$ for a given total cost C,
(II) minimise the total cost C, for a given value of $\mathrm{Var}(\bar{x}_{st})$.

I. *Minimum variance for fixed cost.* We must choose $n_1, n_2, \ldots, n_k$ to minimise

$$\mathrm{Var}(\bar{x}_{st}) = \sum_{i=1}^{k} W_i^2 S_i^2 / n_i - \frac{1}{N} \sum_{i=1}^{k} W_i S_i^2 \qquad \text{(see (4.2))}$$

subject to the constraint

$$\sum_{i=1}^{k} c_i n_i = C - c_0.$$

Introducing a Lagrangian multiplier*, λ, we will need

$$\frac{-W_i^2 S_i^2}{n_i^2} + \lambda c_i = 0, \qquad (i = 1, \ldots, k),$$

* Readers unfamiliar with this technique may either accept the results (*4.13*) and (*4.14*) on trust, or may demonstrate them by the (lengthier) type of substitution argument used on page 27.

D

or

$$n_i\sqrt{\lambda} = W_i S_i/\sqrt{c_i}.$$

Multiplying each side by c_i and summing over the strata, gives

$$(C-c_0)\sqrt{\lambda} = \sum_{i=1}^{k} W_i S_i \sqrt{c_i}.$$

Thus, the *optimum allocation for fixed total cost* is given by

$$n_i = \frac{(C-c_0)W_i S_i/\sqrt{c_i}}{\sum_{i=1}^{k} W_i S_i \sqrt{c_i}} \qquad (4.13)$$

and the *total sample size* will be

$$n = \frac{(C-c_0)\sum_{i=1}^{k} W_i S_i/\sqrt{c_i}}{\sum_{i=1}^{k} W_i S_i \sqrt{c_i}} \qquad (4.14)$$

We see, then, that the stratum sample sizes will need to be proportional to the stratum size, proportional to the stratum standard deviation, and inversely proportional to the square root of the unit sampling cost in the stratum. (Large, highly variable strata with low unit sampling costs will lead to large samples relative to other strata.)

A special case of these results arises when the unit sampling costs c_i are the same in all strata, so that

$$C = c_0 + nc,$$

where c is the (constant) unit sampling cost.

The optimum allocation now needs

$$n_i = \frac{W_i S_i}{\sum_{i=1}^{k} W_i S_i} n \qquad (4.15)$$

with

$$n = \frac{C-c_0}{c}. \qquad (4.16)$$

Clearly the allocation (*4.15*) can be equivalently regarded as *optimum for fixed sample size, ignoring variation in costs of sampling from one stratum to another*, in the sense that, given n, it minimises $\text{Var}(\bar{x}_{st})$.

The allocation (*4.15*) is given a special name: it is called *Neyman Allocation*, after J. Neyman who gave an early proof of its optimality. The resulting minimum variance (for Neyman allocation; that is either for fixed sample size ignoring sampling costs, or with a cost limit and constant unit sampling costs) is obtained by substituting (*4.15*) in (*4.2*) as

$$\text{Var}_{min}(\bar{x}_{st}) = \frac{1}{n}\left(\sum_{i=1}^{k} W_i S_i\right)^2 - \frac{1}{N} \sum_{i=1}^{k} W_i S_i^2, \qquad (4.17)$$

the second term arising from the f.p.c. In the cost-limited case, with constant unit sampling costs, n will need to be determined from (*4.16*).

II. *Minimum cost for fixed variance.*   Suppose that, instead of putting a limit on total cost, we fix $\text{Var}(\bar{x}_{st})$—perhaps by imposing some precision requirement in the form that $\bar{x}_{st}$ should have a certain (high) probability of not differing from $\bar{X}$ in absolute value by more than a specified amount. We would now like to satisfy (for a prescribed value of V) a condition

$$\text{Var}(\bar{x}_{st}) = V$$

for the minimum possible total cost. The appropriate allocation of stratum (and total) sample sizes is immediately determined from the results under I. We can see this informally from the following argument. We know that for any specified total cost, $\text{Var}(\bar{x}_{st})$ is minimised when the n_i are chosen to be proportional to $W_i S_i/\sqrt{c_i}$. For given V there will be some total cost C for which this allocation yields V as the minimum variance. If C were made larger, then, in view of the explicit form of n_i given by (*4.13*), the minimised variance would *become less than* V; this we do not require. If C were smaller, by a similar argument we see that no allocation could restrict $\text{Var}(\bar{x}_{st})$ to as small a value as V. Thus choice of n_i proportional to $W_i S_i/\sqrt{c_i}$ *must also minimise total cost for a given value of* $\text{Var}(\bar{x}_{st})$.

Specifically we need

$$n_i = k W_i S_i/\sqrt{c_i},$$

where k must be chosen to ensure that

$$\text{Var}(\bar{x}_{st}) = \sum_{i=1}^{k} W_i^2 S_i^2/n_i - \frac{1}{N} \sum_{i=1}^{k} W_i S_i^2 = V.$$

Hence we must take

$$n_i = \left\{ \frac{\sum\limits_{i=1}^{k} W_i S_i \sqrt{c_i}}{V + \frac{1}{N} \sum\limits_{i=1}^{k} W_i S_i^2} \right\} W_i S_i / \sqrt{c_i}, \qquad (4.18)$$

and the total sample size n will be

$$n = \frac{(\sum\limits_{i=1}^{k} W_i S_i \sqrt{c_i})(\sum\limits_{i=1}^{k} W_i S_i / \sqrt{c_i})}{V + \frac{1}{N} \sum\limits_{i=1}^{k} W_i S_i^2}.$$

Again, if unit sampling costs are constant at some value c, we see that Neyman allocation *(4.15)* is optimum in the sense of *minimising total sample size* (since this is equivalent to minimising total cost) for a given $\text{Var}(\bar{x}_{st}) = V$. The resulting minimum total sample size will be

$$(\sum\limits_{i=1}^{k} W_i S_i)^2 \Big/ \left(V + \frac{1}{N} \sum\limits_{i=1}^{k} W_i S_i^2\right).$$

One more situation warrants investigation. Optimum allocation may not be feasible; suppose we must operate with *prescribed sampling weights*, $w_i = n_i/n$, for the different strata. It is nonetheless useful to examine how to choose the total sample size to achieve a specified value for $\text{Var}(\bar{x}_{st})$.

III. *Sample size needed to yield some specified* $\text{Var}(\bar{x}_{st})$ *with given sampling weight*. Suppose we want $\text{Var}(\bar{x}_{st}) = V$. For example, we might specify a margin of error, d, and acceptable probability of error, α, in the sense that we need

$$Pr\{|\bar{x}_{st} - \bar{X}| > d\} \leqslant \alpha.$$

Proceeding as in Section 2.5, using an assumed normal distribution for $\bar{x}_{st}$ (but see Section 4.4), this would require

$$V = (d/z_\alpha)^2.$$

Equating $\text{Var}(\bar{x}_{st})$ to V gives

$$n = \sum\limits_{i=1}^{k} (W_i^2 S_i^2 / w_i) \Big/ \left(V + \frac{1}{N} \sum\limits_{i=1}^{k} W_i S_i^2\right).$$

Thus, as a first approximation to the required sample size, we have

$$n_0 = \frac{1}{V} \Big(\sum\limits_{i=1}^{k} W_i^2 S_i^2 / w_i \Big),$$

or more accurately

$$n = n_0 \left(1 + \frac{1}{NV} \sum_{i=1}^{k} W_i S_i^2\right)^{-1}.$$

In the special cases of *proportional allocation*, and *Neyman allocation*, we have

$$n_0 = \frac{1}{V} \sum_{i=1}^{k} W_i S_i^2; \qquad n = n_0(1 + n_0/N)^{-1},$$

and

$$n_0 = \frac{1}{V} \left(\sum_{i=1}^{k} W_i S_i\right)^2; \qquad n = n_0 \left(1 + \frac{1}{NV} \sum_{i=1}^{k} W_i S_i^2\right)^{-1},$$

respectively.

4.4 Comparison of Proportional Allocation and Optimum Allocation

One further aspect in studying the efficiency of the stratified sample mean is to ask to what extent optimum allocation of sample units to the different strata is better than proportional allocation. Proportional allocation is straightforward; it requires no knowledge of stratum variances or relative sampling costs. This type of knowledge *is* required for optimum allocation in the sense of the previous section. Such knowledge may not be available, or only known imprecisely. Its acquisition may require fairly detailed preliminary enquiries, or the acceptance of certain assumptions about the population structure that are difficult to justify. Before embarking on such enquiries, formulating such assumptions, or considering the implications of an imprecise statement of sampling costs or stratum variances, we should have some idea of what *potential* gain can arise from optimum, rather than proportional, allocation.

We shall consider just one case, the comparison of proportional allocation and Neyman allocation (optimum for constant unit sampling costs in different strata).

Denoting $\text{Var}(\bar{x}_{st})$ by V_P and V_N, for proportional and Neyman allocation, respectively, we must have $V_P \geqq V_N$, and from (4.4) and (4.17),

$$V_P - V_N = \frac{1}{n} \sum_{i=1}^{k} W_i(S_i - \bar{S})^2,$$

where $\bar{S} = \sum_{i=1}^{k} (W_i S_i)$. Thus the extent of the potential gain from optimum (Neyman) allocation compared with proportional allocation depends on the *variability of the stratum variances*: the larger this is, the *greater the relative advantage of optimum allocation.*

Example 4.2

Consider once again the situation described in Example 2.4 where we wish to estimate the total Christmas Card sales for a network of 243 stationery shops, by asking some of the shops to submit 'early returns' at the end of January. Suppose that for general accounting purposes the shops have been divided into three groups on the basis of their average annual turnover for all products over the period of the previous five years. The finite population is thus stratified. The full July returns of the Christmas Card sales over recent years enable us to make fairly precise statements of the three stratum variances. It is reasonable to expect that sampling costs will be higher for the larger shops. The strata sizes, variances, and sampling costs (in appropriate units) are as follows:

Average turnover (£'000)	N_i	S_i^2	c_i
Less than 50	146	0·16	2
Between 50 and 100	62	0·58	3
Greater than 100	35	0·31	4

Suppose we again want to estimate total sales for the current year, in such a way that we have a 95% chance of our estimate being within 10% of the true figure. This requires (again assuming that total sales will be in the region of 420, and using the normal approximation) that we restrict the variance of our stratified sample estimator of total sales to about 460. Equivalently, we need $\operatorname{Var}(\bar{x}_{st}) = V = (0{\cdot}0882)^2$.

With proportional allocation we will have sample stratum weights 0·601, 0·255 and 0·144, respectively. Now

$$\frac{1}{V} \sum_{i=1}^{k} W_i S_i^2 = 37{\cdot}126,$$

and

$$37{\cdot}126(1 + 37{\cdot}126/243)^{-1} = 32{\cdot}206.$$

Thus, allowing for the fact that sample sizes must be whole numbers, we will need to take 34 observations: with 20, 9 and 5 observations, respectively, drawn at random from the three strata.

Taking the calculations a stage further we might seek the optimum allocation. The sample weights now need to be about 0·527, 0·348, and 0·124, respectively. The required total sample size is now 31, consisting of 16, 11, and 4 observations from the different strata.

What we should note from these results is the impressive reduction in the required sample size (from 62 to about 33) that arises from exploiting the stratification of the population. (The reduction for optimum, compared with proportional, allocation (from 34 to 31) is modest by comparison). We would expect to obtain such an improvement from stratification in this situation, in view of the stratum variances being small relative to the population variance (which is of the order of 0·64), and in view of the fact that the stratum means will inevitably vary widely (being correlated with total turnover, our basis of stratification).

4.5 Some Practical Considerations

We have considered in some detail the effects, and possible advantages, of stratification of the population in relation to the estimation of $\bar{X}$ and (implicitly) X_T. The discussion has been rather formal. It was assumed that the values of certain population characteristics—such as stratum sizes, N_i, and variances, S_i^2—are known. Little attention was given to real life considerations in the choice of strata, or to the practical problems of determining stratum sample sizes when, as is likely, we do not have very accurate knowledge of the N_i or S_i^2. In this section we shall briefly examine such matters.

Unknown N_i and S_i^2

The various results we have obtained for $\mathrm{Var}(\bar{x}_{st})$ under different circumstances, and for the choice of stratum sample sizes, have been expressed in terms of the N_i and S_i^2. Very often in practice we will have no precise knowledge of the values of these quantities. At best we can estimate them from the survey data, or informally assign 'reasonable' values on the basis of previous experience. Any conclusions must reflect this lack of precise knowledge.

Even if we know the stratum sizes N_i, and adopt some prescribed allocation of stratum sample sizes n_i, then uncertainty of the values of the $S_i{}^2$ will mean that we cannot accurately assess the variance of the estimator $\bar{x}_{st}$ of $\bar{X}$. The estimation of $\text{Var}(\bar{x}_{st})$ from the sample data has been briefly discussed in Section 4.1.

If the N_i are also unknown, greater difficulties arise. The stratum weights $W_i = N_i/N$ are crucial ingredients in $\text{Var}(\bar{x}_{st})$, and if we are to use a stratified sample mean we must estimate them in some way. It is possible that published data may help. For example, nationally based returns such as the Census for human population factors, or other large-scale government department surveys of agricultural, industrial, medical or educational factors, will contain a great deal of breakdown of information for the country as a whole. If it is reasonable to believe that some 'local' population *represents* the larger national environment, then we can adopt the national stratum weights in the local enquiry. But such representativeness is far from inevitable—local populations are notoriously idiosyncratic. The determination of stratum weights in this way *can* be perfectly reasonable—it *can* be fraught with danger. The major safeguard lies in the experience, and shrewdness, of the investigator. Previous successes and failures must guide him in any decision to carry over 'global' stratum weights to the local problem. Statistical methods are of little help, except on those rare occasions where the potential gains from stratification are large enough to justify the expense of some fairly detailed pilot survey *designed purely to estimate the stratum weights* as a preliminary to the main stratified sample survey. This approach is sometimes called *double sampling*, or *two-phase sampling*, and it is possible to take formal account of the sampling properties of the estimators of the W_i (implied by the method of drawing the pilot data) in assessing the properties of $\bar{x}_{st}$.

Certain general remarks can be made concerning the likely effects of imprecise knowledge of the W_i. It will usually happen that $\bar{x}_{st}$ will be biased; its accuracy is best assessed by mean square error about $\bar{X}$, rather than by $\text{Var}(\bar{x}_{st})$. Furthermore, the bias does not tend to reduce with increase in sample size. *Estimation* of the mean square error (necessary because the stratum variances will also be unknown) will introduce further imprecision.

But lack of knowledge of the N_i and $S_i{}^2$ (particularly the latter) becomes even more serious in the matter of choice of the allocation of stratum sample sizes. Such an allocation must be determined *before* we choose the sample, so that we do not even have sample estimates of the S_i^2 to help us. This is the same problem (but in more acute form) as the one considered in Section 2.5 concerning the choice of the size of a s.r. sample in an *unstratified* population. Some of the suggestions made at the earlier stage may again prove fruitful: the use of provisional estimates of variance from pilot studies or from similar surveys carried out previously. Or again we may consider taking preliminary samples from each stratum to yield rough estimates of stratum variances which are in turn used to determine an appropriate allocation of stratum sizes. The preliminary sample is then 'topped up' to the required allocation. This is another illustration of *double* (or *two-phase*) *sampling*. Imprecision arises using any of these methods and will be reflected in, for example, what we might hope to be an optimum allocation being non-optimum in practice—perhaps seriously so. Cochran (1963, Chapter 5A) considers in more detail the effects of non-optimum allocation, also the effects of several other aspects of incomplete knowledge of stratum sizes and variances. See also Sampford (1962, Chapter 6) for further practical details.

Over-sampling of strata

When applying the formulae (*4.13*) or (*4.18*) to determine optimum stratum sample sizes, it is not impossible that some of the resulting n_i may *exceed* the corresponding stratum sizes N_i. This is particularly likely to happen if the sampling fraction is large and if stratum variances differ widely. Obviously we cannot take more observations that there are members in a stratum, and the optimum allocation cannot be attained. The usual practice is to *fully sample* any strata for which the optimum n_i exceed N_i—that is, to take all members of those strata. For these strata $\bar{X}_i$ and S_i^2 will be determined exactly. The variance of the resulting $\bar{x}_{st}$ cannot now be obtained from the results for optimum allocation, and we must take care to use the appropriate form of the general expression (*4.2*) (or its estimator (*4.6*)), recognising that certain f_i will be unity and hence that the corresponding terms in $\mathrm{Var}(\bar{x}_{st})$ will not contribute.

Sub-populations; several variables; multi-way stratification

These represent just a few of the possible extensions of study of stratified populations, and only brief comments will be made.

D*

In a sample survey we often wish to estimate characteristics of *sub-populations* (or '*domains of study*') as well as characteristics of the total population. For example, in a survey of beef prices at wholesale meat markets throughout the country, *regional* differences might be of interest. Different parts of the sample data will have been drawn from different regions and will constitute a basis for estimating regional characteristics. Then again, national average prices for different *cuts* of meat may be a principal concern—the data again provides access to these. In stratified sampling, we can distinguish two different possibilities. The sub-populations, or domains, may themselves constitute the basis for stratification, in which case estimation in sub-populations is particularly simple. Recalling that stratification is most advantageous when stratum means differ widely, and within-stratum variation is low, then stratifica-tion by different cuts of beef will clearly be better than stratification by geographic region. Suppose the population has been stratified on such a basis. A stratified sample in this situation consists of s.r. samples *for each cut* of meat, and therefore s.r. sample means can be used to estimate mean prices for each cut and we can immediately assess the precision of the estimates.

But a word of caution is needed. If the *overall* mean (say) is of little interest, and the major concern is for the estimation and inter-comparison of sub-population (here *stratum*) means, then the earlier results on opti-mum allocation of stratum sample sizes may not apply. For example, it might be more appropriate to optimise with respect to precision of some estimator of contrast between stratum means. Alternatively, if the regional means (over all cuts) are of interest, we have to recognise that the sample data for each region *spans all the strata*. For any region we will *not* have a s.r. sample since the sample size is a *random* quantity. Appropriate estimators, and their variances, are now more complicated and would require special investigation. This difficulty is not restricted to stratified sampling, of course.

Any large scale survey is bound to be costly. We will inevitably wish to extract the maximum amount of useful information from it. This commonly implies taking simultaneous observations on *several* variables, X, Y, Z etc., for each sampling unit. Why, for example, restrict attention to beef prices? It is a simple matter to include lamb prices, or even to extend our enquiries to wider aspects of the operation of the meat markets such as different measures of their size (for instance, their total sales over a day, or the size of the geographic area they serve). One difficulty that now arises in stratified sampling is that what constitutes an optimum (or even a reasonable) allocation of stratum sizes for one variable may not do so for another. The first stage must be to determine and compare appropriate allocations for the different variables separately. Often they will not differ widely, because of similarities and correlations between the variables (beef and lamb prices), and a compromise can be adopted which is not far from the appropriate (optimum or, perhaps, proportional) allocation for each. If they do differ widely there can be no sensible compromise, and additional criteria must be introduced. These will usually be based on an assessment of the relative importance of the different variables; this might be a purely subjective procedure or occasionally a more formal decision theory type of analysis might be conducted.

In large scale surveys there will often be great appeal in stratifying the population with respect to several factors simultaneously. In a sociological enquiry it might appear desirable to divide the population into different sexes, different employment groups, different nationalities, different types of accommodation, and so on. Several practical interests support this: a desire to make the sample 'reflect' the population as a whole, or to facilitate the study of highly specialised subgroups (possibly different groups are of particular relevance for different variables being simultaneously recorded). The *strata* become correspondingly specialised: female, highly paid, Welsh, lathe operators living in high-rise flats, and so on! For both the estimation of overall population characteristics, and the characteristics of subgroups, there might be some disadvantages in such *multi-way stratification*. The individual factors of stratification (sex, income group, etc.) will have been chosen for practical interest. They may not necessarily correspond to desirable bases of stratification in the statistical sense of leading to improved efficiency of estimation of overall population characteristics. Simultaneous stratification by several factors soon leads to a vast number of strata—just 3 factors each at 4 levels yields $4^3 = 64$ strata. Except in a very large survey, individual stratum sample sizes are bound to be very small. The precision of estimators of stratum means (totals, etc.) will be correspondingly low. Administrative difficulties are likely to arise in the choice of random samples for the specialised subgroups. These subgroup (stratum) sizes and variances are unlikely to be known (or to be estimable) with any precision, so that the appropriate allocation of stratum sample sizes, or the estimation of the variance of estimators, will be difficult and unreliable. One particular problem is that there may be so many strata that it is not economical to ensure that they are all represented even with a single observation. Such a problem is discussed and illustrated by Cochran (1963, Chapter 5A), who once again provides a more detailed development of the various topics introduced throughout this sub-section.

Post-hoc stratification

Suppose plans have been drawn up to conduct a sample survey on a stratified population, and that stratum sizes and stratum variances are known. We can use the results derived in the earlier sections of this chapter to determine the variance of the stratified mean $\bar{x}_{st}$ for any allocation of stratum sample sizes—proportional allocation might be a typical choice. Alternatively we could calculate the optimum allocation that should be used to minimise $\text{Var}(\bar{x}_{st})$ or total cost. But when all is said and done, the success of our efforts will rest on our ability to actually obtain the appropriate sample in the practical situation. We considered earlier some of the practical difficulties of drawing a truly random sample. In stratified sampling, we need such a sample *from each stratum*, and a major complication can arise in the respect that *we may not be able to determine in which stratum an observation belongs, until it has been drawn*. This can happen, for example, where strata correspond to different personal details on people—such as their religious beliefs, income levels, nationality, educational achievement, and so on. For such factors, published national reports may provide a clear indication of stratum weights (sizes) and variances; but it can be most difficult to sample individuals from specific strata. We may be forced to draw our sample and stratify it subsequently: that is, carry out a *post-hoc stratification*.

If this problem exists there can be no prospect of drawing a stratified random sample, since we cannot draw s.r. samples from specific strata. One possibility is to take a s.r. sample of size n from the whole population, and subsequently assign individuals to the different strata. Although we do not obtain a stratified random sample, we should expect to encounter numbers in the different strata roughly in proportion to the stratum sizes, N_i. The resulting stratified sample should be somewhat similar to that obtained by stratified random sampling with proportional allocation, provided that numbers of individuals, n_i', falling in each stratum, i, are reasonably large. If we were now to estimate $\bar{X}$ by the quantity analogous to $\bar{x}_{st}$ (rather than by $\bar{x}$), that is by

$$\tilde{\bar{x}} = \sum_{i=1}^{k} W_i \tilde{\bar{x}}_i,$$

where $\tilde{\bar{x}}_i$ is the mean of the observations that are in stratum i, we might expect to recover some of the potential advantages of $\bar{x}_{st}$ itself.

This proves to be so; as long as the S_i^2 do not differ widely, and the n_i' are large, $\tilde{\bar{x}}$ behaves similarly to $\bar{x}_{st}$ obtained from proportional allocation. Its variance is thus approximately $(1-f)\Sigma W_i S_i^2/n$, which (as we have seen) can be considerably less than $\mathrm{Var}(\bar{x}) = (1-f)S^2/n$.

Another possible use of post-hoc stratification is to correct 'obvious lack of representativeness' in a s.r. sample.

Example 4.3

Suppose we draw a random sample of 10 individuals from the *Class Example* data, and wish to estimate the population mean height $\bar{X}$. The particular sample drawn happens to contain 7 women, and 3 men; the sample mean is

$$\bar{x} = 6\cdot8 .$$

But we know that 60% of the population are men; our sample contains only 30% men and $\bar{x}$ is surely likely to under-estimate $\bar{X}$ (which is 8·44). If instead of using $\bar{x}$, we work out the sample means for the men and women separately, we obtain, respectively,

$$\tilde{\bar{x}}_M = 11\cdot00, \qquad \tilde{\bar{x}}_F = 5\cdot00 .$$

The weighted estimator $\tilde{\bar{x}}$ for such post hoc stratification yields

$$\tilde{\bar{x}} = 0\cdot6 \times 11\cdot00 + 0\cdot4 \times 5\cdot00$$

$$= 8\cdot6$$

which is a much more reasonable value.

Clearly, little can be claimed (other than vague intuitive appeal) for such a procedure in such small samples, or without some prescription of what degree of 'lack of representativeness' will be needed to prompt the use of $\tilde{\bar{x}}$ rather than $\bar{x}$. But if the sample size is large enough to be confident that stratum sample sizes will also be large, and $\tilde{\bar{x}}$ is always used, whatever the constitution of the sample, we return to the earlier situation where $\tilde{\bar{x}}$ has essentially similar properties to $\bar{x}_{st}$ obtained by proportional allocation.

Quota Sampling

Closely allied to stratified random sample is the method of *Quota Sampling*, which is widely used in market research, opinion surveys, and a variety of nationwide enquiries. It is a principal method of sampling employed by commercial data collection organisations who service business needs for survey information, as well as being commonly used by individual agencies requiring regular re-appraisal of attitudes or activities in society. Political views, reactions to new or proposed government policy, patterns of trade and industry, consumer attitudes to products, and television audience sizes are all likely to be assessed through surveys based on quota sampling principles. The ubiquitous 'opinion poll' provides full scope for the method.

In essence, quota sampling is merely stratified random sampling with a complex multifactor stratification and stratum sample sizes drawn by proportional allocation. The strata are chosen principally to ensure a 'representative picture' of the population with respect to the factors of stratification, and to yield estimates in specific subgroups, rather than in a desire to enhance efficiency of estimation in a statistical sense. But this latter 'spin off' can arise if the strata happen to have appropriate form (wide discrepancy of mean values, low internal scatter). The utilitarian interest may often produce this effect—consider, for example, stratification by age, employment group, geographic region, etc., in different situations.

Where quota sampling differs from stratified random sampling is in the fact that *the stratum samples may not be random*; an element of subjective choice enters into the sampling practice because of the manner in which it is conducted. Typically, strata are defined, and the sample stratum sizes needed for proportional allocation are then calculated from (more or less) known overall stratum sizes in the population. The data are collected by instructing interviewers or interrogaters to fill the *quotas* for the different strata, by street interviews, house to house enquiries, postal questionnaires, and so on. The selection of respondents is not a random one. To fill the quota by arbitrary selection would be time consuming. Successively more and more observations would be rejected as time goes on as quotas fill up, and the practice is adopted of allowing the interviewer to 'use his judgment' to fill the quotas. Thus as time goes on he exercises more and more personal choice in picking respondents—even style of dress can have a large influence. To fill a quota of 'over 50 year old, professional class men', the local doctor may well be deliberately omitted from a street interview if he has called at the shops to replenish paint stocks, in the midst of painting the attic. This element of 'determined choice' means that we cannot be confident in applying the results above in the quota sampling context. In particular, non-response which is inevitably ignored in Quota Sampling can seriously bias results; consider (for example) television or radio audience assessments, where non-response might be highly correlated with viewing or listening behaviour.

This is not to say that quota sampling cannot produce very good results. It can, and often does. The difficulty is that we have no proper basis for measuring their propriety, since the sampling scheme is not truly probability based. (See Section 1.5). Stephan and McCarthy (1958) give a detailed appraisal of quota sampling methods and practice.

4.6 Estimating Proportions

The earlier development of the properties of stratified sample estimators was restricted to the estimation of the population mean, $\bar{X}$, by the stratified sample mean, $\bar{x}_{st}$. The results extend in an obvious way to the estimation of the population total, X_T. Even within the limited range of study of this text we should take the enquiries a little further. Two obvious omissions concern the estimation of a population proportion, and the use of an auxiliary associated variable to improve the efficiency of estimation of $\bar{X}$ or X_T. A few results on these two topics will be derived in this section and the following one, respectively.

Suppose that each member of a finite population can be assigned to one or other of two categories, and that P is the proportion of the population in the first category. It is of interest to examine how P might be estimated from a stratified random sample, and what the properties of the estimator might be. In a sense, the results derived for $\bar{x}_{st}$ answer this question: P can be regarded as the population mean value of a variable Y, which is zero if population members fall into the second category, one if in the first category. Thus $\bar{Y} = P$, and the corresponding stratified sample mean $\bar{y}_{st}$ seems a sensible choice of estimator for P.

We have

$$\bar{y}_{st} = \frac{1}{n} \sum_{i=1}^{k} W_i \bar{y}_i,$$

with $\bar{y}_i$ being just the proportion p_i of members of the ith stratum sample in the first category of classification. So, we can estimate P by

$$p_{st} = \frac{1}{n} \sum_{i=1}^{k} \frac{N_i}{N} p_i = \frac{1}{n} \sum_{i=1}^{k} W_i p_i, \qquad (4.19)$$

and such a weighted average has an obvious intuitive appeal.

Recalling the discussion of Section 2.7 which considered the estimation of P from a s.r. sample, the only essential distinction between the properties of p_{st} and $\bar{x}_{st}$ will arise from the fact that the stratum variances S_i^2 must now depend on the quantities P_i, the true stratum proportions. We have

$$S_i^2 = \frac{N_i}{N_i - 1} P_i(1 - P_i).$$

Clearly

$$\boxed{p_{st} \text{ is unbiased for } P}$$

and, from (4.2),

$$\boxed{\operatorname{Var}(p_{st}) = \sum_{i=1}^{k} \frac{W_i^2}{n_i} \left(\frac{N_i - n_i}{N_i - 1} \right) P_i(1 - P_i)} \qquad (4.20)$$

If the N_i are at all large, then the factors $(N_i - n_i)/(N_i - 1)$ can of course be replaced by $(1 - f_i)$.

Special forms of this result for different allocations $(n_1, ..., n_k)$, and optimum choice of allocation, all follow from the results in Sections 4.1–4.4.

In particular, with *proportional allocation*, we have

$$\operatorname{Var}(p_{st}) = \frac{N-n}{n} \sum_{i=1}^{k} \frac{W_i^2}{(N_i - 1)} P_i(1 - P_i),$$

$$\doteq \frac{(1-f)}{n} \sum_{i=1}^{k} W_i P_i(1 - P_i),$$

(replacing $(N_i - 1)$ in the denominators by N_i).

Adopting this latter approximation, the optimum allocations *for fixed sample size ignoring costs (Neyman allocation)* and *for fixed cost*, $C = c_0 + \sum_{i=1}^{k} (c_i n_i)$, are

$$n_i = \frac{n W_i \sqrt{[P_i(1 - P_i)]}}{\sum_{i=1}^{k} W_i \sqrt{[P_i(1 - P_i)]}},$$

and

$$n_i = \frac{(C-c_0)W_i\sqrt{[P_i(1-P_i)/c_i]}}{\sum\limits_{i=1}^{k} W_i\sqrt{[P_i(1-P_i)c_i]}},$$

respectively. (See Section 4.3.) The potential advantages of such optimum allocations over arbitrary or proportional allocation can be assessed from the appropriate forms of the results in Section 4.4.

In practice we will not know the P_i, and must use appropriate sample estimates. For example, we will estimate $S_i{}^2$ by the unbiased estimator $n_i p_i(1-p_i)/(n_i-1)$. (See Section 2.7.)

4.7 Ratio Estimators in Stratified Populations.

In Chapter 3 we considered some of the ways in which the existence of an auxiliary variable Y, 'correlated' with the variable X of principal interest, can be exploited to provide better estimators of characteristics of the population of X values than are obtained from a s.r. sample of X values alone. The *ratio estimator*, or *regression estimator*, provided this facility. One or the other was the more appropriate choice depending on the nature of the apparent relationship between the X and Y values (and depending on the extent of their association). If these estimators are to be recommended in simple random sampling from unstratified populations, there is reason to believe that the same may be true in stratified populations. To illustrate this we shall consider just one of the ways in which an associated auxiliary variable Y can be employed in estimating characteristics of the population of X values. The example we shall take is that of a *ratio estimator* of $\bar{X}$ in a stratified population.

Suppose we draw a stratified random sample of values of (X, Y), with $n_1, \ldots, n_k$ observations in the different strata, the sample stratum means being, $\bar{x}_i, \bar{y}_i (i = 1, \ldots, k)$. Suppose that in looking at the scatter diagrams of the data for each stratum there appears to be a fair degree of proportionality between the values of the two variables, shown in the form of a roughly linear relationship through the origin without substantial scatter. The slope need not appear to be identical in each stratum. Such an indication suggests that ratio estimators of stratum means or totals is likely to be profitable. (See Section 3.2.) We must assume, of course, that the stratum mean values $\bar{Y}_i$, for the auxiliary variable, are known.

There are various ways in which we can combine such ratio estimators for the different strata to yield an estimator for the whole population. Two possibilities are to use the so-called *separate ratio estimator*, or *combined ratio estimator*. Consider estimating $\bar{X}$. The ratio estimator of $\bar{X}_i$ is $(\bar{x}_i/\bar{y}_i)\bar{Y}_i$, where $\bar{x}_i$, $\bar{y}_i$ are the stratum sample means.

Then the *separate ratio estimator* of $\bar{X}$ is

$$\bar{x}_{sst} = \sum_{i=1}^{k} W_i \frac{\bar{x}_i}{\bar{y}_i} \bar{Y}_i,$$

i.e. the weighted average of the separate ratio estimators.

The *combined ratio estimator* reverses this process, forming first the stratified sample means and then correcting for the relationship between the two variables. It has the form

$$\bar{x}_{cst} = \frac{\bar{x}_{st}}{\bar{y}_{st}} \bar{Y}.$$

The corresponding estimators of the total X_T will be

$$\sum_{i=1}^{k} W_i \frac{\bar{x}_i}{\bar{y}_i} Y_{iT}$$

(Y_{iT} is the total of the Y values in the ith stratum), and

$$\frac{\bar{x}_{st}}{\bar{y}_{st}} Y_T,$$

respectively.

What of the merits of these estimators? Both will tend to be biased unless the sample size is reasonably large. The bias will be less serious in $\bar{x}_{cst}$ than in $\bar{x}_{sst}$ since the sample size condition applies to the total sample rather than to each stratum sample. Approximate variances of $\bar{x}_{sst}$ and $\bar{x}_{cst}$ can be obtained by combining the results for the stratified mean and ratio estimator (typically (*4.2*) and (*3.10*)). It turns out that, unless the relationship between X and Y is the same in all strata, the *separate* estimator will be more 'efficient' than the *combined* estimator. But this must be offset by the lower tendency to bias in $\bar{x}_{cst}$, and the fact that we do *not* need to know the separate stratum means $\bar{Y}_i$, only the overall mean $\bar{Y}$.

The combined effect of stratification and the use of ratio estimators is somewhat unpredictable. It is appealing to think that potential gains from stratification should be further enhanced by using ratio estimators. But it is not as simple as this: the two effects are often tied up. Stratification can have the effect of reducing (even annihilating) the potential advantage of the relationship between X and Y, by weakening the relationship within the strata. The effects are by no means additive and the best we can say is that by combining the two techniques we should do as well as the better of the two on its own, for the problem in hand. We can sometimes do much better; particularly when the auxiliary variable does *not* serve as the basis for stratification. Coupled with the extra effort, and knowledge, necessary if $\bar{x}_{sst}$ or $\bar{x}_{cst}$ are to be employed, this makes their use problematical.

Needless to say all the separate practical and formal difficulties of stratified sampling, and ratio estimation, will enter into their use in combination (including problems of estimating variances, etc.). In terms of survey *design* the form of optimum allocation will tend to be somewhat different to that described in Section 4.3 when the ratio method is used. See Cochran (1963, Section 6.14).

Analogous methods exist for using regression estimators in stratified populations, where these are more appropriate than ratio estimators.

Example 4.4

The separate ratio estimator $\bar{x}_{sst}$ has been calculated for 500 stratified random samples of size 5 from the *Class Example* data, with proportional allocation, and rows as strata. The histogram of values so obtained is shown in

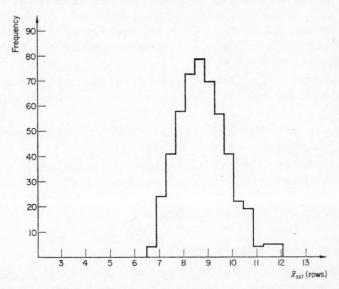

Figure 9 Histogram of 500 values of $\bar{x}_{sst}$ (rows) for samples of size 5 in the *Class Ex mple*.

Figure 9; the mean and variance of the 500 values of $\bar{x}_{sst}$ are 8·82 and 1·02, respectively. In Example 4.1 we saw the dramatic effect of stratification by rows for estimating $\bar{X}$. Taking account also of the relationship between heights and weights in the population, by using $\bar{x}_{sst}$, turns out to be far less successful. The variance of the estimator is increased, and there is serious bias ($\bar{X} = 8·44$).

4.8 Conclusions

This chapter has covered a large amount of material to do with sampling and estimation in stratified populations. To set the results in perspective it is desirable that we conclude the chapter with a review of some of their qualitative features. This is conveniently achieved by posing, and briefly answering, a few questions.

Why Use Stratified Populations?

(i) In the hope of obtaining more efficient estimators than would be possible without stratification.

(ii) For administrative convenience; practical constraints of access or cost may compel different sampling techniques to be used for different sections of the population. The resulting data arise as random samples from the different sections; these constitute the strata in what is a stratified random sample.

(iii) Because of an intrinsic interest in the sub-populations corresponding to the strata; or a desire to 'represent' such sub-populations 'fairly'.

(iv) To reduce fortuitous bias in an unstratified sample, by post-hoc stratification—but this has dubious utility.

Individually, or in combination, these factors support the use of stratified populations, either by deliberate construction of appropriate strata (as in (i) and (iv)) or in an inevitable form determined by practical constraints and interests (as in (ii) and (iii)).

What are the Advantages?

As implied by (i), (ii), and (iii), we would hope to obtain increased efficiency of estimation of population characteristics under appropriate circumstances, additional sampling convenience or greater ease of access to sub-populations of special interest.

When does Stratification lead to Improved Efficiency?

Population characteristics can be more efficiently estimated from a stratified sample than from an overall s.r. sample if *strata means differ widely, and within-strata variation is low.* The greater this effect of stratification the greater the efficiency of the corresponding estimators. With freedom of choice of strata, the aim should be to construct strata with these characteristics. If stratification is largely for administrative convenience the choice is limited, the efficiency improvement uncertain (although practical constraints do often produce a subdivision of the population appropriate to improvement of efficiency).

How should the Population be Stratified?

If unhampered by practical constraints, then clearly the aim should be to divide the population into non-overlapping groups of X values to maximise the separation, and internal homogeneity, of the strata. So the proper basis for stratification to achieve maximum efficiency is the set of X values itself. In practice, however, we will not have sufficient knowledge of the population to stratify it in this way. Instead we must employ some more tangible external criterion. If, as often happens, this corresponds reasonably well with separation of the X values into non-overlapping groups, little potential advantage from stratification will be lost. For example, stratification by sex should prove a good criterion when estimating measures of physical stature, stratification by geographic region, or occupational category, likewise in estimating socio-economic factors. Although inevitably tempered by sampling ease and cost and the knowledge of stratum sizes and variances, we should strive to stratify the population in a way that is likely to produce the sort of stratification required from the theoretical standpoint. Note in particular that nothing is really achieved in terms of efficiency of estimation by stratifying a population in such a way that the different strata are essentially indistinguishable statistically.

How should the Sample Sizes be allocated to different Strata?

Proportional allocation is particularly straightforward, and will often extract most of the potential advantages of stratified sampling. It is commonly used. If reliable information on stratum variances and on sampling costs is available, then optimum allocation (or for constant unit sampling costs, Neyman allocation) is to be recommended.

Exercises for Chapter 4

I

A survey is to be conducted to estimate the total number of books borrowed from the 217 public libraries in a county authority during a particular week. It is possible to classify the libraries as small, medium, and large in size, on the basis of their stocks of books. The numbers of books borrowed from libraries in the three groups are thought to be roughly in the proportions $1:2:3$. It is further anticipated that the within-group variances of numbers of books borrowed will be proportional to the square root of the corresponding means. There are 71 small, 126 medium, and 20 large libraries.

A total sample size of about 40 is required. If sampling costs in each group are the same, how should sample sizes in a stratified random sample be allocated to the three groups?

II

A stratified population has 5 strata. The stratum sizes, N_i, and means and variances, $\bar{X}_i$ and S_i^2, of some variable X are as follows.

Stratum	N_i	$\bar{X}_i$	S_i^2
1	117	7·3	1·31
2	98	6·9	2·03
3	74	11·2	1·13
4	41	9·1	1·96
5	45	9·6	1·74

Calculate the overall population mean and variance, $\bar{X}$ and S^2. For a stratified random sample of size about 80, determine the appropriate stratum sample sizes under proportional allocation, and Neyman allocation. Work out (for the same total size of sample) the efficiency of the s.r. sample mean $\bar{x}$ as an estimator of $\bar{X}$, relative to the stratified sample means for the two methods of allocation.

III

A stratified population of total size N is made up of k strata of sizes $N_1, N_2, \dots, N_k$. The ith stratum contains a proportion P_i of members possessing a particular characteristic $(i = 1, 2, \dots, k)$. If the f.p.c. can be ignored, show that the variance of the stratified random sample estimator of the overall proportion, P, of population members possessing the particular characteristic is approximately

$$\frac{1}{n}\left\{ \sum_{i=1}^{k} \frac{N_i}{N} \sqrt{[P_i(1-P_i)]} \right\}^2$$

when the stratum sample sizes are optimally allocated.

For $k = 2$, compare the efficiencies of the stratified random sample estimators of P for equal allocation, proportional allocation, and optimum allocation of stratum sample sizes. Illustrate the way in which the *minimum* relative efficiencies for equal, or proportional, allocation (compared with optimum allocation) change with different values of the P_i, for a representative range of such values.

IV

In a stratified population the sampling cost for obtaining a stratified random sample of size n, made up of n_i observations from the ith stratum $(i = 1, 2, \ldots, k)$, is

$$C = c_0 + \sum_{i=1}^{k} c_i \sqrt{n_i}.$$

The stratified sample mean, $\bar{x}_{st}$, is to be used to estimate the population mean $\bar{X}$.

Determine the optimum allocation of the n_i (to minimise $\text{Var}(\bar{x}_{st})$ for fixed total cost C).

V

Dairy farms in a certain geographic region are divided into four categories, depending on their total acreage and on whether or not they concentrate exclusively on dairy products. The numbers of farms in the four categories are 72, 37, 50, and 11. In a survey to estimate the total number of milk-producing cows in the region, a stratified random sample of 28 farms is chosen with (roughly) proportional allocation. The numbers of cows on the selected farms are:

Category	Number of cows
1	61, 47, 44, 70, 28, 39, 51, 52, 101, 49, 54, 71
2	160, 148, 89, 139, 142, 93
3	26, 19, 21, 34, 28, 15, 20, 24
4	17, 11

Estimate the total number of cows on farms in these 4 categories throughout the region and the standard error of the estimator.

5 Cluster Sampling

It is convenient at this stage to recall some basic distinctions that were drawn in the opening chapter: the differences between the *target population*, the *study population*, and the *sampling frame* in a sample survey. The target population is the population about which we are seeking information. The study population is the population from which we actually draw our sample data. Ideally the two should coincide. In practice they sometimes do not do so, and care must be exercised to ensure that any conclusions drawn from the sample data, and thus relating to the study population, can legitimately be extrapolated to the target population. For example, in attempting to assess the attitudes of the adult population of the country (the *target population*) to certain political or social issues, it may be necessary to seek information by interviewing people in the streets. The population of people in the streets, from whom the sample is drawn, is the *study population*. Can it be assumed to represent the attitudes of the total adult population? Clearly the populations do not fully coincide; for a variety of reasons certain people may not be accessible 'in the streets', also regional or other socio-economic differences in the target population may not be fully represented in the study population unless great care is taken to do this.

The *sampling frame* is the formal representation of the study population which serves as the basis for choice of members of the population. It may take the form of an actual list of population members from which our sample can be chosen 'on paper' according to the prevailing sampling scheme; the chosen sample of population members must then be sought 'in the field' and the required information elicited. But such a procedure cannot always be followed, and many technical difficulties arise. Even with a full list of all members of the study population, on which we can specify the chosen sample, it can be difficult or costly to seek out the members of the sample. To try to obtain a purely random sample in a nationwide enquiry will involve a great deal of travelling or correspondence, and possibly loss of information due to non-response or failure to find chosen sample members.

Frequently the sampling frame does not coincide precisely with the study population or target population. For example, we are unlikely to

possess a full list of all adult members of the population. Street interviewing is an informal means of overcoming the non-existence of a
detailed list of population members. Fortuitous encounter is hoped to
yield a 'random sample' from the study population of people 'in the
streets'; such people are hoped to be 'representative' of the adult
population at large.

But many survey enquiries fall between the two extremes of a fully
listed, easily accessible population, and a haphazard choice from a
restricted study population chosen principally on grounds of accessibility. The sampling frame often exists for a division of the target (or
study) population into non-overlapping *groups* of population members.
All population members are represented in these groups, which may be
of different sizes. There may be a convenient list of the groups in the
population, which can be used for specifying the sample that will be
sought. The sampling frame thus provides a coverage of the population
of interest, but its members (the *sampling units*) do not correspond to
individual members of the population. Such loss of identification of
individuals is offset by the convenience of having a tangible list of
sampling units in which to define a sample and, possibly, by practical
advantages of cost or access in contacting chosen *sample* members
(which are of course sampling units, or *groups* of members of the
population). For example, a list of addresses might be a convenient
basis of access to individuals—but each address may correspond to
several people. Or in an enquiry into the performance of schoolchildren,
choice of a sample of local educational authorities and study of schools
administered by those authorities is likely to be easier and less expensive
than choice of schoolchildren individually (irrespective of whether or
not a complete list of schoolchildren exists).

Note how such dual considerations of the convenience of a list for
specifying a required sample, and cost and access advantages, imply
that we essentially sample from a *stratified population*. The strata are the
sampling units, which are likely to be many in number, each containing
relatively few population members. But methods of stratified sampling
are unlikely to be appropriate. As described in Chapter 4 these involve
sampling *from each stratum*. The administrative stimuli for using a
sampling frame with units representing many small strata imply
selection of such strata. Thus instead of selecting some population
members within *each* stratum, we will wish to select *some* strata but
possibly study *each* selected stratum in full. This difference of emphasis
is reflected in different terminology. The strata are now called *clusters*;

the choice of a sample of such clusters to yield a sample of the population members is called *cluster sampling*. If all the population members in each selected cluster are used in the sample, the method is known as *one-stage cluster sampling*. If not, but further selection is exercised within the chosen clusters, the technique is called *sub-sampling*, or *two-stage cluster sampling*. Sometimes the clusters themselves consist of groups of population members. The clusters are then called *primary units*, their constituent sub-groups *secondary units*, and so on. Selection of a sample by choosing from among the primary units, from secondary units within chosen primary units, and so on to further stages, is called *multistage cluster sampling*. For instance, in examining performance of primary schoolchildren in the educational survey referred to above, local education authorities may constitute the primary units, their schools the secondary units, and children within the schools the members of the study population.

In the discussion of *stratified sampling* in Chapter 4, it was recognised that the manner in which a population is stratified may be conditioned by administrative factors. Nonetheless, the major interest in stratification is in its potential value for producing more efficient estimators of population characteristics. In contrast, *cluster sampling* is employed almost exclusively for administrative convenience; either to ease sample specification through the existence of a list of the clusters, or to improve access to the population, or to reduce sampling costs. Cluster sampling methods are, and need to be, widely employed. Often other methods, were they feasible, would produce more efficient estimators but at greater cost and administrative effort.

The hope is that any loss in potential efficiency is outweighed by reduction in sampling costs, and greater sampling facility. Any objective comparison of cluster sampling with other methods needs to be made on this basis: it is unrealistic to exclude cost considerations in such a comparison, although the accurate specification of costs is not always easy, and it would be optimistic to claim that cluster sampling in practice is often supported by a justificatory cost analysis. Pragmatism is the major stimulus.

5.1 One-stage Cluster Sampling with Equal Sized Clusters

As in our study of other sampling methods, we shall again restrict attention to just one or two factors of cluster sampling. In particular, only estimation of a population mean $\bar{X}$, total X_T, or proportion P, will be considered, and discussion of the comparison (or combination) of cluster sampling with other methods will be limited. No theoretical or practical details of complicated multi-stage sampling are given, nor any study of two-phase (or multi-phase) sampling (see Section 4.5) for prior estimation of cluster variances or correlations. Neither is there space to take up the issue of how to choose the basis of clustering, when several possibilities exist.

Suppose that the population consists of a set of *clusters* of individual population members. There are M clusters of sizes $N_1, N_2, ..., N_M$ ($\Sigma_{i=1}^M N_i = N$). The members of the clusters are X_{ij} ($i = 1, 2, ..., M$; $j = 1, 2, ..., N_i$), and the *cluster means* and *cluster variances* are $\bar{X}_i$ and S_i^2 ($i = 1, 2, ..., M$), defined in the usual way. The population mean and variance are $\bar{X}$ and S^2, respectively. A sample of population members is obtained by taking a *simple random sample* of m clusters and including in the sample *all* members of the chosen clusters. The resulting sample, of size $n \geq m$, is a *one-stage cluster sample*. How are we to use it to estimate $\bar{X}$? The simplest case to study is where all clusters are of the same size.

Suppose
$$N_1 = N_2 = ... = N_M = L, \quad \text{say.}$$
Then
$$N = ML.$$

The one-stage cluster sample, arising as a s.r. sample of m clusters, has size
$$n = mL;$$
the sampling fraction is
$$f = n/N = m/M.$$

Suppose the observations in the sample are x_{ij} ($i = 1, 2, ..., m$; $j = 1, 2, ...L$).

As an estimator of $\bar{X}$, the *cluster sample mean* might be considered. This is just
$$\bar{x}_{cl} = \frac{1}{mL} \sum_{i=1}^m \sum_{j=1}^L x_{ij}. \qquad (5.1)$$

Since all clusters are of the same size, $\bar{x}_{cl}$ is clearly *unbiased* for $\bar{X}$; that is

$$\boxed{E(\bar{x}_{cl}) = \bar{X}}$$

Furthermore its variance is easily shown to be

$$\text{Var}(\bar{x}_{cl}) = \frac{1}{m}(1-f)\sum_{i=1}^{M}\frac{(\bar{X}_i - \bar{X})^2}{M-1}. \qquad (5.2)$$

These results arise from the fact that $\bar{x}_{cl}$ can be expressed as

$$\frac{1}{m}\sum_{i=1}^{m}\bar{x}_i,$$

where the $\bar{x}_i$ are the cluster means for a s.r. sample of m of the M clusters. Regarding the set of cluster means $\{\bar{X}_i, \bar{X}_2, \ldots, \bar{X}_M\}$ as the basic population from which we are sampling, this population has mean

$$\frac{1}{M}\sum_{i=1}^{M}\bar{X}_i = \bar{X}$$

and variance

$$\frac{1}{M-1}\sum_{i=1}^{M}(\bar{X}_i - \bar{X})^2.$$

The unbiasedness of $\bar{x}_{cl}$ as an estimator of $\bar{X}$, and the form (5.2) for its variance, now follow directly from the earlier results (see (2.3)) for a s.r. sample mean.

Consider the alternative estimator of $\bar{X}$ provided by the mean, $\bar{x}$, of a s.r. sample of $n(= mL)$ observations drawn from the total population (ignoring the cluster structure). This is also unbiased, and has variance $(1-f)S^2/(mL)$. How does this compare in efficiency with $\bar{x}_{cl}$?

Note that

$$
\begin{aligned}
(ML-1)S^2 &= \sum_{i=1}^{M}\sum_{j=1}^{L}(X_{ij} - \bar{X})^2 \\
&= \sum_{i=1}^{M}\sum_{j=1}^{L}(X_{ij} - \bar{X}_i)^2 + L\sum_{i=1}^{M}(\bar{X}_i - \bar{X})^2 \\
&= M(L-1)\bar{S}^2 + L\sum_{i=1}^{M}(\bar{X}_i - \bar{X})^2, \qquad (5.3)
\end{aligned}
$$

where

$$\bar{S}^2 = \frac{1}{M}\sum_{i=1}^{M}S_i^2$$

is the *average within-cluster variance*.

Thus

$$\text{Var}(\bar{x}) - \text{Var}(\bar{x}_{cl}) = \frac{1-f}{mL(M-1)} \{(M-1)S^2 - L \sum_{i=1}^{M} (\bar{X}_i - \bar{X})^2\}$$

$$= \frac{(1-f)M(L-1)}{mL(M-1)} (\bar{S}^2 - S^2), \qquad (5.4)$$

by (5.3).

So the question of the relative efficiency of $\bar{x}$ and $\bar{x}_{cl}$ (in terms of a straight comparison of their variances, ignoring cost or convenience factors) has a simple resolution. *The cluster sample mean, $\bar{x}_{cl}$, will be better than the s.r. sample mean, $\bar{x}$, if the average within-cluster variance, $\bar{S}^2$, is larger than the overall population variance, S^2; and vice versa.*

It is interesting to observe that this is the reverse of what was found in stratified sampling, where the method yielded greater efficiency if the within-strata variation was sufficiently *low*. Bearing in mind the basic difference in the two sampling techniques (cluster sampling and stratified sampling), this reversal is what would be expected intuitively!

There is another way of looking at these results. We can define a quantity called the *intra-cluster correlation coefficient*,

$$\rho = 2 \sum_{i=1}^{M} \sum_{j<k} (X_{ij} - \bar{X})(X_{ik} - \bar{X}) / [(L-1)(ML-1)S^2], \qquad (5.5)$$

which provides an aggregate measure of the correlation between population members in the same cluster. This clearly has affinities with $\bar{S}^2$; the larger the value of ρ, the smaller we would expect $\bar{S}^2$ to be in general. In fact,

$$\rho = 1 - \left(\frac{ML}{ML-1}\right)\left(\frac{\bar{S}^2}{S^2}\right)$$

(see Exercise 2 of this chapter),

and we can express $\text{Var}(\bar{x}_{cl})$ as

$$\text{Var}(\bar{x}_{cl}) = \frac{(1-f)}{m} \frac{ML-1}{L^2(M-1)} S^2[1+(L-1)\rho]. \qquad (5.6)$$

The condition, $\bar{S}^2 > S^2$, for $\bar{x}_{cl}$ to be more efficient than $\bar{x}$, now becomes

$$\rho < -\frac{1}{(ML-1)} \sim 0,$$

so that, as long as the population is large, the requirement for greater efficiency of $\bar{x}_{cl}$ is that the *intra-class correlation should be negative*. (Again, see Exercise 2 of this chapter.)

It should be recognised that, since the prime stimulus for using a cluster sample is one of convenience, the greater efficiency of $\bar{x}_{cl}$ over $\bar{x}$, although feasible, is not likely to be widely encountered.

Estimation of the population total X_T is, of course, effected by $ML\bar{x}_{cl}$, and its variance is merely the appropriate multiple of (5.2).

In practice we will need to *estimate* $\mathrm{Var}(\bar{x}_{cl})$, and an unbiased estimator is obtained by replacing $\Sigma_{i=1}^{M}(\bar{X}_i-\bar{X})^2/(M-1)$ in (5.2) by $\Sigma_{i=1}^{m}(\bar{x}_i-\bar{x}_{cl})^2/(m-1)$. Under appropriate conditions we can again construct approximate confidence intervals for $\bar{X}$, or X_T, by assuming that $\bar{x}_{cl}$ is normally distributed.

Example 5.1

Consider, once again, the data on heights of students given in the *Class Example*. We could consider either (i) the rows, or (ii) the columns, as 5 equal-sized clusters in the population. Picking one row, or one column, at random yields a cluster sample of size 5. To estimate the mean height $\bar{X}$ the cluster sample mean might be considered in either case. Note that only 5 possible values can arise (in each case) and they do so with equal probabilities. The two sampling distributions are thus particularly simple. They do not need to be estimated by taking a large number of samples, as was done previously to illustrate results for s.r. sampling, ratio estimation, or stratified sampling. The possible values for $\bar{x}_{cl}$ (and the within-cluster variances) are (see Example 4.1) in case

(i)   3·4,   7·2,   9·2,   11·8,   10·6,
      (3·3), (1·2), (2·2), (6·2), (7·3),

(ii)   9·0,   8·0,   7·2,   8·4,   9·6,
      (10·5), (14·0) (6·7), (25·3), (13·8).

The average within-cluster variances are 4·04 and 14·06 respectively. The exact sampling distributions are presented in Figure 10. The exact variances of $\bar{x}_{cl}$ are easily obtained by working out the variances of the possible $\bar{x}_{cl}$ values in cases

(i) and (ii) (that is by using the result (5.2)). We obtain, in case

(i) $\text{Var}(\bar{x}_{cl}) = 8 \cdot 69$,

(ii) $\text{Var}(\bar{x}_{cl}) = 0 \cdot 68$.

The variance of a s.r. sample mean, $\bar{x}$, based on a s.r. sample of size 5 from the whole population, was found to be 1·99. Thus we confirm the results above for cluster sampling. When the average within-cluster variance is smaller than S^2 ($= 12 \cdot 42$), $\bar{x}$ is better than $\bar{x}_{cl}$. This is markedly so in case (i). When it is larger than S^2, $\bar{x}_{cl}$ is better than $\bar{x}$. This happens in case (ii).

We note the expected contrast with stratified sampling, where stratification by rows produced a large improvement in efficiency over $\bar{x}$; by columns, not so!

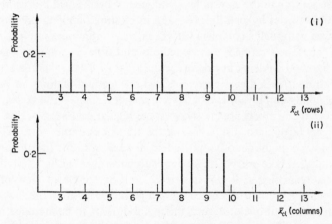

Figure 10 Sampling distributions of $\bar{x}_{cl}$ for (i) rows, (ii) columns, for samples of size 5 in the *Class Example*.

The same principles apply for estimating the population proportion, P, of X values satisfying some criterion. Suppose P_i ($i = 1, 2, ..., M$) are the corresponding cluster proportions. Our sample yields m of these; $p_1, p_2, ... p_m$. The cluster sampling estimate of P is now

$$p_{cl} = \frac{1}{m} \sum_{i=1}^{m} p_i,$$

which is just a special case of $\bar{x}_{cl}$, where X is assigned new values 1 or 0 depending on whether or not it satisfies the criterion of interest. The variance of p_{cl} is obtained from (5.2) by replacing the $\bar{X}_i$, and $\bar{X}$, by the P_i, and P, respectively.

Before proceeding to discuss what happens when the clusters have different sizes, we shall consider a special form of cluster sampling which is very widely used in practice.

5.2 Systematic Sampling

Suppose we wish to draw a sample of size n from a population of size N and have available a complete list of the population members. The list is of great value in specifying the sample we intend to draw. Members of the sample will be identified on the list, and then sought in the population. Thus we might sample the students in a university by using a published list of students for that university; or the books in a library by going through the card index catalogue of books held by the library (multiple entries in such indexes raise interesting sampling problems!). But even with such a complete list, the choice of random sample (simple even, let alone stratified) can be tedious and time-consuming. Imagine choosing 500 students *at random* from a list of 8500. There is a great temptation to seek an easy way out, and one method of doing so, which is commonly employed, is to take a *systematic sample*. The principle is that sample members are chosen in some regular manner working progressively through the list. Consider the student example; since $8500 = 500 \times 17$, we might choose any student *at random* in the first 17 on the list, and then take every 17th student subsequently. This is a systematic sample. Note that it is not a strictly *random* sample in view of the deterministic method of choice of sample members after the first one. If N is not a multiple of n, then the method needs to be modified in an obvious way. The sampling is clearly very easy to carry out, which is a great advantage, particularly in frequently repeated surveys, or when the sample is chosen 'on site' rather than prior to the collection of the data. Sometimes even the limited randomisation, in the choice of the first member, is dispensed with, and the sequence is fully prescribed at the outset. For example, it might be decided to take the 9th student, and every 17th one subsequently, on the basis that this seems to be a 'symmetric' policy. Or to save even more time in the library example we might choose cards by taking cards out every one inch (say) through the index—thus obtaining a 'more or less' systematic sample.

Apart from the saving in time and effort, there is also some sort of intuitive appeal in systematic sampling: it seems to 'span the population' in a way that might lead to more 'representative' results than those obtained from random choice. Is this appeal justified? To examine this we will consider a systematic sample chosen by taking one member at random from the first M on the list, and every Mth subsequent one. Suppose this yields a sample of size n. There are two ways of viewing such a situation. The population can be thought of as divided into strata consisting of the first M members, the second M members, and so on. The sample is a correspondingly stratified one with precisely one member from each stratum. But it is not a stratified *random* sample, since the observations are not chosen *at random* in each stratum. Thus the results of Chapter 4 will not immediately apply. Alternatively, and more usefully, we can view the sample as a *cluster sample*. The population consists of M clusters:

$$X_1, X_{M+1}, X_{2M+1} \cdots$$
$$X_2, X_{M+2}, X_{2M+2} \cdots$$
$$\vdots$$
$$X_M, X_{2M}, X_{3M} \cdots ,$$

of sizes differing by at most one. The sample is now a cluster sample, consisting of one cluster chosen at random from the M clusters. Suppose $N = LM$, so that all clusters have the same size $L = (N/M) = n$; we can immediately apply the results of Section 5.1, with $m = 1$. If N/n is not an integer, so that the cluster sizes are not all exactly the same, minor modifications will be necessary in the terms of the next section, but the effects are qualitatively unaltered.

To estimate the population mean $\bar{X}$ we take the *systematic sample mean*

$$\bar{x}_s = \frac{1}{n} \sum_{i=1}^{n} x_i,$$

where $x_1, x_2, \ldots, x_n$ are the observations in the single chosen cluster of the population. In the notation of Section 5.1, $\bar{x}_s$ is just $\bar{x}_{c1}$ based on $m(= 1)$ clusters chosen at random from the $M = N/L$ 'systematic' clusters of size $L = n$ into which the population has been divided. So $m = 1, L = n, M = N/n, f = n/N$. We conclude that

$$\boxed{E(\bar{x}_s) = \bar{X}}$$

E

so that $\bar{x}_s$ is *unbiased*, and by (5.2),

$$\mathrm{Var}(\bar{x}_s) = \frac{1}{M}\sum_{i=1}^{M}(\bar{X}_i - \bar{X})^2 \qquad (5.7)$$

$$= \frac{n}{N}\sum_{i=1}^{N/n}(\bar{X}_i - \bar{X})^2,$$

where $\bar{X}_i$ is the mean of the ith cluster (or the ith of the M *potential* systematic samples). Equation (5.3) becomes

$$(N-1)S^2 = M(n-1)\bar{S}^2 + Mn\,\mathrm{Var}(\bar{x}_s),$$

so that if $\bar{x}$ is the mean of a s.r. sample of size n

$$\mathrm{Var}(\bar{x}) - \mathrm{Var}(\bar{x}_s) = \frac{n-1}{n}(\bar{S}^2 - S^2).$$

The systematic sample mean will be more efficient than $\bar{x}$ if $\bar{S}^2 > S^2$; that is, in the event that the average variance within systematic samples is larger than the overall population variance. This can sometimes happen, and endorses the intuitive feeling that systematic sampling (as well as being easy to carry out) can be statistically advantageous if systematic division of the population results in widely differing X values in each potential systematic sample. In practical terms this effect depends on the way in which the population has been listed (the listing may be highly structured in terms of X values, or at the other extreme essentially random). Cochran (1963, Chapter 8) considers this matter in some detail.

The results can again be expressed in terms of the intra-cluster correlation ρ. We have, from (5.6), that $\bar{x}_s$ has

$$\mathrm{Var}(\bar{x}_s) = \left(1 - \frac{1}{N}\right)[1 + (n-1)\rho]S^2/n,$$

and efficiency in excess of $\bar{x}$ if

$$\rho < -1/(N-1) \sim 0,$$

that is (essentially) if ρ is negative.

If N/n is not an integer, some potential systematic samples will have one more member than others: in this respect the clusters in the population are not now all of the same size. The effect of such (small) differences in cluster sizes will be qualitatively unimportant unless the systematic sample size is very small.

5.3 One-stage Cluster Sampling with Different Sized Clusters

Suppose as before that the population consists of M clusters, but that their sizes are $N_1, N_2, \ldots, N_M$ ($\Sigma_{i=1}^{M} N_i = N$) where not all the N_i have the same value. A cluster sample is again drawn as the basis for estimating some population characteristic, say $\bar{X}$. The cluster sample consists of *all* members of each of m clusters randomly selected from the M clusters in the population. Suppose that the sizes, means, and totals of the *chosen* clusters are n_i, $\bar{x}_i$, x_{iT} ($i = 1, 2, \ldots, m$). Certain complications now arise because the cluster sizes differ; various alternative estimators of $\bar{X}$ might be considered, and their sampling behaviour is not always easy to determine precisely. We shall consider here just three possible methods of estimating $\bar{X}$ (with obvious extensions to the estimation of X_T or P); one further possibility is discussed later (Section 6.3).

It is useful to distinguish between the *primary units* (the clusters) and the *secondary units* (the population members within the clusters). The cluster sample is a s.r. sample of the primary units, and we can carry over the results for s.r. sampling given in earlier chapters to study its behaviour. We are essentially sampling a population of size M, where each member is represented by certain variables: for example, the cluster total X_{iT}, the cluster mean $\bar{X}_i$, or the cluster size N_i; for ($i = 1, 2, \ldots, M$). The cluster sample provides observed values x_{iT}, $\bar{x}_i$, n_i ($i = 1, 2, \ldots, m$) of these variables.

The *overall* population mean $\bar{X}$ (that is the mean value of X over the secondary units) is

$$\bar{X} = \sum_{i=1}^{M} \sum_{j=1}^{N_i} X_{ij} / \sum_{i=1}^{M} N_i = \sum_{i=1}^{M} X_{iT} / \sum_{i=1}^{M} N_i, \qquad (5.8)$$

which is interpretable in the 'reduced' population of primary units as a *population ratio* of the total of the X_{iT} values to the total of the N_i values. If we know both the number of clusters M, and the total overall population size $N = \Sigma_{i=1}^{M} N_i$, then writing

$$\bar{X} = \left(\frac{M}{N}\right) \sum_{i=1}^{M} X_{iT} / M, \qquad (5.9)$$

$\bar{X}$ is alternatively represented as just a known multiple of the primary population mean value of X_{iT} (i.e. of the *mean cluster total* $\bar{X}_T = (1/M) \Sigma_{i=1}^{M} X_{iT}$).

These representations suggest two possible ways of estimating $\bar{X}$ from the cluster sample.

(a) The cluster sample ratio.   Use of the results of Section 3.1 suggests estimating $\bar{X}$ by

$$\bar{x}_{C(a)} = \sum_{i=1}^{m} x_{iT} \bigg/ \sum_{i=1}^{m} n_i, \qquad (5.10)$$

which is the *ratio* of the sum of the cluster totals to the sum of the cluster sizes, in the chosen sample of clusters.

Thus (see Section 3.1) $\bar{x}_{C(a)}$ will have bias of order m^{-1} which will be unimportant only if the number of clusters in the sample is large. The variance of $\bar{x}_{C(a)}$ will be given by the approximation (see (3.1)).

$$\operatorname{Var}(\bar{x}_{C(a)}) \doteqdot \frac{(M-m)M}{(M-1)m} \sum_{i=1}^{M} \left(\frac{N_i}{N}\right)^2 (\bar{X}_i - \bar{X})^2. \qquad (5.11)$$

This variance depends on the variation between the cluster means: the smaller the variation, the smaller the variance. The effect is similar to what was found in the case of equal sized clusters; compare (5.11) and (5.2). $\operatorname{Var}(\bar{x}_{C(a)})$ can be estimated from the sample by

$$\frac{(M-m)M}{m(m-1)} \sum_{i=1}^{m} \left(\frac{n_i}{N}\right)^2 (\bar{x}_i - \bar{x}_{C(a)})^2,$$

and if N is unknown, it may be replaced by the sample estimate Mn/m to yield

$$\frac{(M-m)m}{M(m-1)} \sum_{i=1}^{m} \left(\frac{n_i}{n}\right)^2 (\bar{x}_i - \bar{x}_{C(a)})^2.$$

To estimate the overall total X_T we merely use $N\bar{x}_{C(a)}$; but here knowledge of the value of N is essential. The variance is just $N^2 \operatorname{Var}(\bar{x}_{C(a)})$.

(b) The cluster sample total.   If the total population size is known, (5.9) suggests that we estimate $\bar{X}$ from the usual s.r. sample estimate of a population mean. This implies using the estimator

$$\bar{x}_{C(b)} = \frac{M}{Nm} \sum_{i=1}^{m} x_{iT}. \qquad (5.12)$$

Clearly,

$$E(\bar{x}_{C(b)}) = \frac{M}{N} \bar{X}_T = \bar{X},$$

so that $\bar{x}_{C(b)}$ has the advantage of being strictly *unbiased*.

Its variance is, from (*2.3*),

$$\text{Var}(\bar{x}_{C(b)}) = \frac{(M-m)M}{(M-1)mN^2} \sum_{i=1}^{M} (X_{iT} - \bar{X}_T)^2. \qquad (5.13)$$

Since $\bar{X}_T = (N/M)\bar{X}$, (*5.13*) differs from (*5.11*) merely in the fact that the sums of squares of the cluster totals is calculated about the fixed quantity $M\bar{X}/N$ rather than about the individual $N_i\bar{X}$ which depend on the specific cluster sizes. This implies that (*5.13*) will tend to be larger than (*5.11*), since cluster totals are most likely to be positively correlated with cluster sizes. This possible disadvantage needs to be set against the attraction of the unbiasedness of $\bar{x}_{C(b)}$. But the greater efficiency of $\bar{x}_{C(a)}$ is far from guaranteed: many factors enter into the comparison, including the relationship (if any) between the $\bar{X}_i$ and N_i, and the variability of the cluster sizes. If the N_i do not vary greatly, then $\text{Var}(\bar{x}_{C(a)})$ need not be much different to $\text{Var}(\bar{x}_{C(b)})$ (indeed for equal sized clusters they are identical).

The estimation of $\text{Var}(\bar{x}_{C(b)})$ from the sample data, and corresponding results for estimating the population total, follow in the obvious way. Again there is the advantage that N need not be known for estimating X_T.

Yet another estimator of $\bar{X}$, which has as its principal attraction a very simple form which is easily calculated, is obtained as:

(c) The unweighted average of the chosen cluster means. That is, we use

$$\bar{x}_{C(c)} = \frac{1}{m} \sum_{i=1}^{m} \bar{x}_i. \qquad (5.14)$$

This estimator is biased and inconsistent (in the finite population sense) unless all cluster sizes are the same. It provides a useful 'quick estimate' which will not be too seriously biased unless the cluster means and cluster sizes are highly correlated. Its variance is again obtained from (*2.3*), as

$$\text{Var}(\bar{x}_{C(c)}) = \frac{M-m}{mM(M-1)} \sum_{i=1}^{M} (\bar{X}_i - \bar{X}_C)^2, \qquad (5.15)$$

(where $\bar{X}_C = (1/M) \sum_{i=1}^{M} \bar{X}_i$), which can be estimated from the sample by

$$\frac{M-m}{mM(m-1)} \sum_{i=1}^{m} (\bar{x}_i - \bar{x}_{C(c)})^2.$$

(If the bias is substantial, however, then the expected mean square error can of course be much larger than (5.15)).

The expected value of $\bar{x}_{C(c)}$ is $\bar{X}_C$, so that its bias is

$$\bar{X}_C - \bar{X} = \frac{1}{M} \sum_{i=1}^{M} \bar{X}_i - \frac{1}{N} \sum_{i=1}^{M} N_i \bar{X}_i.$$

If the X_i, or the N_i, do not vary too much, this bias will not be serious, and the estimator can compare reasonably in efficiency with $\bar{x}_{C(a)}$ or $\bar{x}_{C(b)}$.

The corresponding estimator of X_T is $N\bar{x}_{C(c)}$, with variance $N^2 \operatorname{Var}(\bar{x}_{C(c)})$.

5.4 Multi-stage Sampling

There is a variety of ways in which the cluster sampling method may be modified or extended to cope with the specific demands of more complicated situations. One example arises where the selected clusters in the primary cluster sample are themselves *sampled*, rather than fully inspected. Thus if we choose a s.r. sample of m of the M clusters (primary units) which comprise the population, we may then take s.r. samples of sizes $n_1, n_2, \ldots, n_m$ of the secondary elements in the chosen clusters. Typically we have $n_i < N_i$ $(i = 1, 2, \ldots, m)$, in contrast to the sampling scheme of the previous section where $n_i = N_i$. The process of drawing samples from the selected clusters is called *sub-sampling*: the resulting total sample of size $n = \sum_{i=1}^{m} n_i$ is called a *two-stage cluster sample*.

If the secondary units are the individual members of the study population, there is no point in going further. If instead they consist of *groups* of population members, then we might either use all their constituent members, or proceed to further stages of sub-sampling. In this latter case we encounter *multi-stage cluster samples*, corresponding to progressively higher levels of sub-sampling.

As described, the probability sampling mechanism at each stage is simple random sampling.

Such multi-stage sampling can be illustrated for the primary school survey referred to in the opening pages of this chapter. In seeking a sample of primary school children, it could be particularly convenient to regard local education authorities as the primary units, schools under their control as the secondary units, and children in those schools as tertiary units. A *three-stage cluster sample* (a s.r. sample of education authorities, a s.r. sample of schools under each of the selected authorities, a s.r. sample of children in each of the selected schools) has the advantages of being relatively easy to obtain, and of appearing to 'cover the population in a representative manner'. In contrast, without a complete listing of schoolchildren it would be a clumsy procedure to seek a direct s.r. sample from the whole population, whilst limited financial resources could imply that very few authorities (perhaps only one) could be chosen in a one-stage cluster sample, with the resulting risk of serious regional idiosyncracies.

Thus the prime stimulus for multi-stage sampling is again administrative convenience, although scope exists for taking into account variance and cost considerations in the specification of what sizes of sample to take at the different stages.

Only a simple illustration of multi-stage sampling will be considered here; discussion of more complicated structures and more sophisticated probability sampling schemes can be found in the various texts on sampling theory which appear in the Bibliography (particularly Cochran (1963); Hansen, Hurwitz, and Madow (1953); and Raj (1968)).

Suppose we have a population consisting of M clusters, each of similar size L, and we draw a *two-stage cluster sample* by taking l members at random from each of a s.r. sample of m clusters. We assume that the cluster elements are the individual members of the population, and we wish to estimate the mean value $\bar{X}$ of some measure X defined on these members.

The total sample size is $n = ml$; the sample members are x_{ij} ($i = 1, 2, \ldots, m; j = 1, 2, \ldots, l$); the within-cluster sample means are denoted $\bar{x}_i$ ($i = 1, 2, \ldots, m$). We denote by X_{ij} ($i = 1, 2, \ldots, M; j = 1, 2, \ldots, L$), $\bar{X}_i$ ($i = 1, 2, \ldots, M$), and $\bar{X}$, respectively, population members, the cluster means, and the overall population mean.

The simplest estimator of $\bar{X}$ is the analogue of the one-stage cluster sample mean $\bar{x}_{cl}$ (see (5.1)), namely

$$\bar{x}'_{cl} = \frac{1}{ml} \sum_{i=1}^{m} \sum_{j=1}^{l} x_{ij} = \frac{1}{m} \sum_{i=1}^{m} \bar{x}_i. \tag{5.16}$$

It is readily confirmed that $\bar{x}'_{cl}$ is *unbiased* for $\bar{X}$, and that it has variance

$$\mathrm{Var}(\bar{x}'_{cl}) = \frac{M-m}{mM} \sum_{i=1}^{M} \frac{(\bar{X}_i - \bar{X})^2}{M-1} + \frac{L-l}{mLl} \sum_{i=1}^{M} \sum_{j=1}^{L} \frac{(X_{ij} - \bar{X}_i)^2}{M(L-1)} \tag{5.17}$$

$$= \mathrm{Var}(\bar{x}_{cl}) + \frac{L-l}{mLl} \bar{S}^2, \tag{5.18}$$

where $\bar{S}^2$ is again the *average within-cluster variance*. As required, $\text{Var}(\bar{x}'_{cl})$ reduces to $\text{Var}(\bar{x}_{cl})$ if $l = L$, that is for one-stage cluster sampling (complete inspection of each selected cluster). When $l < L$, the variance is increased by the amount $(L-l)\bar{S}^2/mLl$, a contribution which arises from the further sampling variation due to subsampling the selected strata. But care must be exercised in interpreting (5.18)! Whilst it literally declares that $\bar{x}'_{cl}$ has larger variance than $\bar{x}_{cl}$, this latter quantity requires full inspection of all m selected clusters so that the sample size is mL. But $\bar{x}'_{cl}$ is based on a sample of size ml which is typically *less* than mL. The fact that a smaller sample yields an estimator with larger variance is no surprise—in itself it does not reflect on the relative efficiency of estimation of $\bar{X}$ in one- and two-stage cluster sampling.

The proofs of the unbiasedness of $\bar{x}'_{cl}$, and of the form (5.17) for its variance, are quite straightforward using conditional expectation arguments: taking within-cluster expectations conditional on the selected clusters followed by the marginal expectation with respect to the s.r. choice of clusters at the first stage. The details will not be presented.

The population total, X_T, can be estimated by $N\bar{x}'_{cl}$. It is unbiased and has variance $N^2 \text{Var}(\bar{x}'_{cl})$.

In practice we will need to estimate $\text{Var}(\bar{x}'_{cl})$. We can obtain an unbiased estimator as

$$s^2(\bar{x}'_{cl}) = \frac{M-m}{Mm} \sum_{i=1}^{m} \frac{(\bar{x}_i - \bar{x}'_{cl})^2}{m-1}$$

$$+ \frac{L-l}{MLl} \sum_{i=1}^{m} \sum_{j=1}^{l} \frac{(x_{ij} - \bar{x}_i)^2}{m(l-1)} . \tag{5.19}$$

At first sight, the second term in (5.19) contrasts strangely with the second term in (5.17): the divisor m in (5.17) has been replaced by M. The reason for this is that with incomplete inspection of the selected clusters,

$$\sum_{i=1}^{m} \frac{(\bar{x}_i - \bar{x}'_{cl})^2}{m-1}$$

is *not* an unbiased estimator of

$$\sum_{i=1}^{M} \frac{(\bar{X}_i - \bar{X})^2}{M-1} ,$$

and a compensation is needed which changes the divisor in the second term of (5.19) in the way described.

One case where this estimated variance (*5.19*) is particularly easy to obtain is where only a small proportion of the clusters is sampled. This is often the case in complex populations. Then the second term becomes negligible and we can estimate $\text{Var}(\bar{x}'_{cl})$ simply from the values of the sample means of the selected clusters, as

$$\frac{M-m}{Mm} \sum_{i=1}^{m} \frac{(\bar{x}_i - \bar{x}'_{cl})^2}{m-1} \, .$$

Whilst the decision to take a two-stage cluster sample is usually dictated by administrative considerations, some choice can often be exercised in the values of m and l. Within any limitations imposed by the total sample size, or total cost of sampling, different values of m and l will yield estimators with different variances, and the question of optimum choice arises. If substantial sampling costs occur only at the sub-sampling stage, and are the same in all clusters, then optimum choice of m and l for fixed total cost amounts to choosing m and l to minimise $\text{Var}(\bar{x}'_{cl})$ for a prescribed total sample size ml. But such a cost structure is not often realistic. When primary sampling and sub-sampling both involve costs, a more complicated cost model is needed. The simplest way in which such differential costs may be included is through a model which declares that there is a basic overhead cost d_0, that each selected primary unit costs an additional amount d_1, and each secondary unit an increment d_2, so that the total cost of sampling is

$$C = d_0 + d_1 m + d_2 ml. \qquad (5.20)$$

Optimum choice of m and l now requires $\text{Var}(\bar{x}'_{cl})$ to be minimised subject to the constraint (*5.20*) in which C is the total amount of money available for sampling. As in our study of stratified sampling, the dual problem of minimising the total cost of a prescribed $\text{Var}(\bar{x}'_{cl})$ does not require separate study. The optimum value of l is the same in both situations, being approximately

$$\sqrt{(d_1/d_2)} \bar{S} (S_W^2 - \bar{S}^2/M)^{-\frac{1}{2}},$$

where $S_W^2 = \Sigma_{i=1}^{M} (\bar{X}_i - \bar{X})^2/(M-1)$. The corresponding value of m may be obtained, as appropriate, from the prescribed value of C or $\text{Var}(\bar{x}'_{cl})$.

In practice, the slight departures from the optimum value of l which are likely to arise from the inevitable uncertainty of the values of $d_1, d_2,$ $\bar{S}^2$, and S_W^2 are likely to lead to little loss of precision relative to the truly optimum choice. When $S_W^2 - \bar{S}^2/M$ is sufficiently small, optimum

E*

choice of l will be L, so that *one-stage* cluster sampling is indicated as the best policy. This makes sense intuitively, since the variance of the cluster means, $S_W{}^2$, is now much the same as $\bar{S}^2/M$, with the implication that elements are assigned to clusters more or less at random. If so, complete inspection of a few clusters should be as efficient as partial inspection of many for the same total size of sample, $n = ml$. At the same time it will, from (5.20), be less costly, since $d_0 + d_2 ml$ remains fixed whilst $d_1 m$ is minimised.

A much more detailed discussion of the properties of estimators and of the optimal choice of primary, and sub-sample, sizes in multi-stage sampling is given by Cochran (1963, Chapters 10 and 11), for this present situation and also for progressively more complicated situations (with unequal cluster sizes, stratified populations, more sophisticated probability sampling schemes).

5.5 Comment

We have observed that the principal reasons for using cluster sampling techniques are administrative, rather than statistical, ones. The lack of a complete listing of population members, and differing problems of access to different groups of population members, may make it laborious, unfeasible, or costly to seek a s.r. sample from the whole population. Very often the population has a hierarchical structure with successive levels containing smaller groups of population members, and access to the different levels is facilitated by the existence of lists of units at each level. The primary school survey was an illustration of such a structure. The most straightforward method of sampling, from the practical viewpoint, is likely to consist of drawing samples at each level: a sample of primary units, sub-samples of secondary units, and so on. In addition to such relative sampling ease, multi-stage cluster sampling can also have the intuitive appeal of seeming to provide a more representative coverage of the highly structured population. Furthermore, the clusters may be of interest in their own right.

Statistical considerations enter with respect to the choice of what probability sampling scheme should be used, and what sizes of sample should be chosen, at the different stages. This choice will depend on the prevailing costs of sampling, but once we proceed beyond simple situations (such as that described in Section 5.4), the appropriate analysis can be highly complex and even intangible, due to inadequate knowledge of the relevant cost and population variability factors.

Even when a complete list of basic population members does exist, or can be compiled, it is often economically undesirable to take a s.r. sample from the whole population, and cluster sampling is to be preferred. For example, the transport and administrative costs of drawing a s.r. sample of schools throughout Scotland (say) could be enormous. If we happen to have agents in 5 towns, who can speedily and cheaply collect the required information on all schools in their towns, the savings in cost for such cluster sampling will be large. It is unlikely that any loss of efficiency of estimation (for similar sizes of sample) could outweigh the economic considerations. Indeed the efficiency loss could be easily remedied by taking a *larger* cluster sample whilst still possibly retaining a substantial cost saving.

Exercises for Chapter 5

I

A professional association publishes a list of its members. In a survey to estimate the average salary of members of the profession, this list is to be used as the basis for selecting a sample of about 5% of the 2640 members of the profession. Discuss the possible advantages and disadvantages of simple random sampling and systematic sampling in the two cases:

(a) where the list is in alphabetical order,
(b) where the list is in order of length of membership of the professional association.

II

Show that in a population consisting of M equal sized clusters, each of size L, the intraclass correlation coefficient, ρ, defined by (5.5), can be expressed in terms of the population variance, S^2, and average within-cluster variance, $\bar{S}^2$, as

$$\rho = 1 - \left(\frac{ML}{ML-1} \right)\left(\frac{\bar{S}^2}{S^2} \right).$$

Confirm the result (5.6) for the variance of the one-stage cluster sample mean, $\bar{x}_{cl}$. Discuss the implied restriction on the range of possible values of ρ, and show that

$$\frac{\mathrm{Var}(\bar{x}_{cl})}{\mathrm{Var}(\bar{x})} \doteqdot 1 + (L-1)\rho.$$

III

A company, which provides its salesmen with cars for company business only, wishes to estimate the average number of miles covered by each car last year. The company operates from 12 branches, and the numbers of cars, N_i, and means and variances ($\bar{X}_i$ and S_i^2) of miles driven last year (in thousands of miles) for each branch, are as follows.

Branch	N_i	$\bar{X}_i$	S_i^2
1	6	24·32	5·07
2	2	27·06	5·53
3	11	27·60	6·24
4	7	28·01	6·59
5	8	27·56	6·21
6	14	29·07	6·12
7	6	32·03	5·97
8	2	28·41	6·01
9	2	28·91	5·74
10	5	25·55	6·78
11	12	28·58	5·87
12	6	27·27	5·38

Suppose that the average mileage is to be estimated by sampling a few branches at random, and using the figures for all the cars at the chosen branches. This is clearly an easier method of sampling than drawing a s.r. sample from the overall population. Compare the efficiencies of the two methods by working out the standard errors of the *unbiased* cluster sample estimator $\bar{x}_{C(b)}$ for a cluster sample of 4 branches, and of the s.r. sample mean for a s.r. sample of 27 cars (this being the average number of cars which would be obtained in a cluster sample of 4 branches).

What can be said of the use of $\bar{x}_{C(a)}$ in this situation?

6 Some Other Probability Sampling Schemes

In this brief study of sampling theory and survey methods, the scope of our enquiries has had to be severely limited. The emphasis has been on basic principles: on the reasons why different types of sampling and methods of estimation make sense in relation to the structure of the population of interest, and on the properties of estimators of basic population characteristics. We have stopped short of investigating in detail the practical aspects of proposed methods and the more complex situations which will inevitably arise involving stratification, multistage sampling, concomitant variables, and cost factors, either individually or in combination. Neither was it feasible to consider multivariate measures, or population characteristics beyond those basic ones of means, totals, or proportions.

The aim has been to present a balanced introduction to the subject and to lay the foundation for further study which may be pursued in more advanced texts, some of which are mentioned in the Bibliography.

But before concluding there is one further limitation implicit in the discussion of the earlier chapters which needs to be explicitly recognised, and at least briefly remedied.

In Section 1.5 the fundamental concept of *probability sampling* was defined and motivated. It was pointed out that any assessment of the propriety of a sampling scheme, or of the properties of resulting estimators, could only be achieved if the sample was drawn by some prescribed probability mechanism. Given a finite population $X_1, X_2, \ldots, X_N$, a probability sampling scheme specifies the size of sample, n, to be drawn and the probabilities π_i with which we will encounter every possible sample, S_i, of such size. *Simple random sampling* was the first scheme considered in detail, in which the S_i were all possible sets of n *distinct* population members ($\binom{N}{n}$ in all) and these arose with equal probabilities $\pi_i = \binom{N}{n}^{-1}$. Such a scheme was seen to be implemented by drawing population members *individually, without replacement*, in such a way that at each stage any of the remaining population members were equally likely to be chosen. The scheme was thus easily specified and readily implemented using a simple sequential procedure for choosing sample members.

Although recognisable structure in the population, or sampling convenience, made it expedient to proceed beyond a s.r. sample of the whole population to stratified samples or cluster samples, the basic sampling mechanism remained that of *simple random sampling*. We merely combined s.r. samples from sub-populations, or took s.r. samples hierarchically. Of the different sampling schemes considered, only systematic sampling fell outside the sphere of s.r. sampling; in the sense that we could not claim that *each* member of the sample was so chosen.

Clearly s.r. sampling has its advantages as we have seen in the results given in the previous chapters. It is easily applied and readily modified to cope with certain types of prevailing structure in the population. It enables stratification, or auxiliary variables, to be exploited to advantage in the sense of producing desirable estimators of a population mean, total, or proportion. But it is by no means the only feasible scheme. Whilst it is the one most widely used in practice, and serves as a most convenient basis for introducing the basic principles of sampling theory, other schemes are used and we should consider briefly one or two of these.

6.1 Sampling with Replacement with Equal Probabilities

The term simple random sampling has been used throughout to describe random sampling *without replacement*. As a result, samples consist of sets of *distinct* members of the population. An obvious alternative would be to allow the possibility of multiple occurrence of population members in the sample. On this principle, there are N^n possible samples of size n. If a sample is chosen in such a way that all such samples have equal probability, N^{-n}, of occurring, the sampling scheme is described as *simple random sampling with replacement*. This again has the advantage of being easy to implement; we have only to draw observations one at a time (until n are obtained) in such a way that each population member $X_1, X_2, \ldots, X_N$ is equally likely to be chosen at each draw. On the other hand it might seem to be somewhat wasteful of effort, in sampling a *finite* population, to allow multiple occurrence. Let us see how this is borne out in estimating the population mean $\bar{X}$.

Suppose $x_1, x_2, \ldots, x_n$ is a s.r. sample of size n drawn *with replacement*. The x_i need not now be distinct population members. Consider the corresponding s.r. sample mean

$$\tilde{x} = \frac{1}{n} \sum_{i=1}^{n} x_i$$

as an estimator of $\bar{X}$. Each x_i is equally likely to be any X_j ($j = 1, 2, ...,$ N) and so has expectation $\bar{X}$.

Thus again

$$\boxed{\tilde{x} \text{ is unbiased for } \bar{X}}$$

But

$$\text{Var}(\tilde{x}) = \frac{1}{n^2} [n \, \text{Var}(x)],$$

where x is a random value from $X_1, X_2, ..., X_n$. So,

$$\text{Var}(\tilde{x}) = \frac{1}{nN} \sum_{i=1}^{N} (X_i - \bar{X})^2 = \frac{(N-1)S^2}{nN}. \tag{6.1}$$

Comparing this with (2.3), the efficiency of $\tilde{x}$ relative to the (without replacement) s.r. sample mean, $\bar{x}$, is

$$\frac{N-n}{N-1},$$

which can represent quite a substantial loss of efficiency unless the sampling fraction n/N is very small.

For sampling without replacement, we noted that an unbiased estimator of S^2 is

$$s^2 = \frac{1}{n-1} \sum_{i=1}^{n} (x_i - \bar{x})^2.$$

With replacement, the corresponding expression

$$\frac{1}{n-1} \sum_{i=1}^{n} (x_i - \tilde{x})^2$$

clearly has expectation

$$\frac{1}{N} \sum_{i=1}^{N} (X_i - \bar{X})^2 = \frac{(N-1)S^2}{N} = \sigma^2, \text{ say,}$$

and σ^2 becomes the more natural measure of population variance in this situation.

Although, on comparisons of this type, s.r. sampling *with replacement* does not compare favourably with s.r. sampling *without replacement*, the general principle of drawing observations *sequentially with replacement* does have some attractions. These arise for probability sampling schemes of a more complicated nature than simple random sampling.

6.2 Sampling with Replacement with Arbitrary Probabilities

Suppose we were to consider some specific probability sampling scheme in which samples $S_1, S_2, \dots$ can arise with probabilities $\pi_1, \pi_2, \dots$. To assess the usefulness of such a scheme we need to overcome *two* difficulties. Firstly, we need a practical procedure which will actually generate the S_i with their prescribed probabilities π_i. This is often a most difficult task since an arbitrary scheme $\{S_i, \pi_i\}$ may not be amenable to any simple form of implementation. An alternative approach is to state a *generating mechanism* for the sample—this will yield a particular set of possible samples each of which arises with a probability determined from the nature of the generating mechanism. For example, choosing observations one at a time at random and without replacement yields samples in accord with the simple random sampling scheme.

A more complicated generating mechanism will similarly yield samples according to some more sophisticated probability sampling scheme, $\{S_i', \pi_i'\}$ say. From the viewpoint of implementing a probability sampling scheme, it is natural that the emphasis should be placed on the method of generating the sample rather than on specifying the scheme, $\{S_i, \pi_i\}$, and then seeking a means of generating it. With such an emphasis we can still encounter a large range of different probability sampling schemes.

The second difficulty concerns the derivation of the properties of estimators based on any probability sampling scheme. If we are to assess the value of the scheme, we must be able to determine, for example, the variances or expected mean square errors of resulting estimators. Again this can involve very complicated calculations, if indeed it is feasible at all. To facilitate this task, a scheme which is defined in terms of sequential generation of successive sample members is again advantageous, especially so when population members are chosen *with replacement*.

It is therefore convenient to consider sampling schemes resulting from the successive choice of population members, with replacement after each choice. A general scheme of this type arises when, at each stage, the different values $X_1, X_2 ..., X_N$ can occur independently with respective probabilities $p_1, p_2, ..., p_N$. Suppose we so choose a sample of size n, the sample values being $x_1, x_2, ..., x_n$.

Again we might estimate $\bar{X}$ by the simple average

$$\tilde{x} = \frac{1}{n} \sum_{i=1}^{n} x_i.$$

The properties of $\tilde{x}$ are easily determined since $x_1, x_2, ..., x_n$ is a random sample from a known probability distribution. We have

$$E(\tilde{x}) = \mu = \sum_{i=1}^{N} p_i X_i$$

and

$$\text{Var}(\tilde{x}) = \frac{1}{n} \left\{ \sum_{i=1}^{N} p_i X_i^2 - \mu^2 \right\}.$$

But μ will *not* equal $\bar{X}$ except in special circumstances, so that $\tilde{x}$ is in general biased and can have large expected mean square error.

The situation is improved if we use a different estimator in which the observations are divided by N times their respective probabilities of occurrence, so that we use

$$\tilde{\tilde{x}} = \frac{1}{n} \sum_{i=1}^{n} (x_i/Nq_i),$$

where q_i is that member of the set $\{p_1, p_2, ..., p_N\}$ corresponding to the population value chosen as the ith member of the sample. *Equivalently,* we are now using the sample mean of a random sample drawn with replacement from a distribution in which values $Y_i = X_i/Np_i$ arise with probabilities p_i $(i = 1, 2, ..., N)$. The mean and variance of this distribution are $\bar{X}$ and

$$\frac{1}{N^2} \sum_{i=1}^{N} X_i^2/p_i - \bar{X}^2,$$

respectively. Thus $\tilde{\tilde{x}}$ is now *unbiased* for $\bar{X}$, with variance

$$\frac{1}{nN^2} \left\{ \sum_{i=1}^{N} (X_i^2/p_i) - (N\bar{X})^2 \right\}. \tag{6.2}$$

This variance can in principle be arbitrarily small; if we put $p_i = X_i/N\bar{X}$, then each sampled value is precisely $\bar{X}$ and $\text{Var}(\tilde{\tilde{x}}) = 0$!

But this is of no practical use since it implies a knowledge of the precise value of $\bar{X}$ in which case we would not need to sample the population!

However, this pathological result does motivate a particular style of sampling which can often be used with advantage. The optimum aim is to sample the population with the probabilities of selection of different members *proportional to their values*, X_i. Without any auxiliary information about the population being sampled, this is unrealistic. But with some knowledge of the population structure, or the facility for sampling an auxiliary variable, Y, correlated with X, some progress in this direction is possible. As examples, we are led to consider in this respect *sampling with replacement* for ratio estimation, or in cluster sampling.

6.3 Sampling with Replacement with Probability Proportional to Size

Sampling with replacement is not a widely applied principle in day-to-day sample survey work, but it can have the advantages of leading to simple (essentially) unbiased estimators of means, totals, etc. with easily estimated variances in certain situations. We shall consider some examples.

Ratio Estimation

Suppose that along with the values X_i of principal interest we can simultaneously measure values Y_i of an auxiliary variable, positively correlated with the X_i. In Chapter 3, Sections 3.1 and 3.2, we considered the estimation of the population ratio, $R = \bar{X}/\bar{Y}$, and the ratio estimator of the population mean $\bar{X}$. The estimator of R having form

$$r = \bar{x}/\bar{y}$$

was discussed, and its properties derived. We can reinterpret r as a *weighted average* of individual observation values $r_i = x_i/y_i$, in the form

$$\sum_{i=1}^{n} (y_i r_i) \Big/ \sum_{i=1}^{n} y_i.$$

The alternative unweighted average

$$r_u = \sum_{i=1}^{n} r_i = \frac{1}{n} \sum_{i=1}^{n} x_i/y_i$$

was considered only briefly; it can have much more serious bias than r. However, the results of the previous section suggest that the unweighted average might be worth considering if we were to choose members of the sample with differing probabilities. To simplify the analysis, suppose that sampling occurs *with replacement*, but with probabilities proportional to the (assumed known) values of Y. Thus each observation is one of the values $R_i = X_i/Y_i$ $(i = 1, 2, ..., N)$ chosen with probabilities $p_i = Y_i/Y_T$ $(i = 1, 2, ..., N)$. We estimate R by

$$r_u = \frac{1}{n} \sum_{i=1}^{n} (x_i/y_i) = \frac{1}{n} \sum_{i=1}^{n} r_i. \qquad (6.3)$$

Since now $E(x_i/y_i) = R$ (for any i), r_u is *unbiased* and has variance

$$\frac{1}{n} \sum_{i=1}^{N} \frac{Y_i}{Y_T} \left(\frac{X_i}{Y_i} - R \right)^2$$

which can be estimated without bias by

$$s^2(r_u) = \frac{1}{n(n-1)} \sum_{i=1}^{n} (r_i - r_u)^2. \qquad (6.4)$$

Correspondingly, we can estimate $\bar{X}$ or X_T by $r_u \bar{Y}$ and $r_u Y_T$, estimating their variances by $\bar{Y}^2 s^2(r_u)$ and $Y_T^2 s^2(r_u)$.

No clear cut comparison is possible of the relative advantages of using weighted averages from equal probability (without replacement) sampling, or unweighted averages from (with replacement) sampling having probabilities proportional to the Y values. Some broad distinctions can be drawn. The latter estimators will be unbiased, but are likely to have larger variance (if only because of the replacement element) and require a detailed knowledge of the Y-values. The former will be biased, though not seriously so in large samples, but will have smaller variance (especially when the R_i values differ widely at *large* Y_i values). When the bias is likely to be substantial and the R_i values differ widely at *small* Y_i values, the with-replacement (probabilities proportional to Y) estimator (6.3) is to be recommended.

The method of sampling with probabilities proportional to Y can be further modified by *inverse sampling* until a required number, n, of distinct population members is obtained, or by sampling *without replacement*. It is more difficult to derive the properties of estimators based on these sampling schemes, although it is known that they can be better than their direct sampling (with-replacement) analogues. See, for example, Sampford (1962, Chapter 7).

Cluster Sampling

Another situation in which samples may usefully be drawn with different probabilities attached to the occurrence of different population members is in the sphere of cluster sampling. This will be illustrated for one-stage sampling although the technique has (perhaps greater) advantages when applied to the choice of the primary units in multi-stage sampling. Suppose it is necessary to estimate the population total, X_T, for a population which consists of M clusters of sizes $N_1, N_2, ..., N_M$ ($\Sigma_{i=1}^{M} N_i = N$) with cluster totals X_{iT} ($i = 1, 2, ..., M$). For this purpose a single stage cluster sample, of $m < M$ clusters, is to be chosen. For simple random sampling the results have been presented in Section 5.3. In many situations we will find that the larger clusters contribute the larger values X_{iT} towards the total X_T. It might seem sensible, therefore, to give greater attention to such clusters, perhaps by giving them larger probabilities of occurrence in the sample.

To keep the analysis fairly simple, we shall consider sampling *with replacement*, in such a way that the ith population cluster (of size N_i; $i = 1, 2, ..., M$) has probability p_i of being chosen. From the above results the ideal value of p_i is X_{iT}/X_T. The corresponding estimator of X_T is then

$$\tilde{\tilde{x}}_T = \frac{N}{m} \sum_{i=1}^{m} \left[x_{iT} \frac{X_T}{N x_{iT}} \right] \equiv X_T, \qquad (6.5)$$

with *zero sampling variance*. But again this is not feasible, since if we knew X_T we would not need to estimate it. However, an associated alternative measure of cluster size *which may be known* is the number of population members, N_i, in the cluster. This suggests sampling the clusters with probabilities $p_i = N_i/N$, and the estimator becomes

$$\tilde{\tilde{x}}_T = \frac{N}{m} \sum_{i=1}^{m} \bar{x}_i, \qquad (6.6)$$

which is just N times the average value of the means of the chosen clusters. Using an argument similar to that preceding (6.2), we see that this estimator is *unbiased*, and has variance

$$\text{Var}(\tilde{\tilde{x}}_T) = \frac{N}{m} \sum_{i=1}^{M} N_i(\bar{X}_i - \bar{X})^2 \qquad (6.7)$$

Corresponding to (6.6) we have an estimator of $\bar{X}$ of the form

$$\tilde{x} = \tilde{x}_T/N = \frac{1}{m} \sum_{i=1}^{m} \bar{x}_i, \tag{6.8}$$

which is *unbiased* and has variance

$$\mathrm{Var}(\tilde{x}) = \mathrm{Var}(\tilde{x}_T)/N^2.$$

We can obtain unbiased estimators of $\mathrm{Var}(\tilde{x}_T)$ and $\mathrm{Var}(\tilde{x})$ from the sample, in the form

$$s^2(x_T) = \frac{N^2}{m(m-1)} \sum_{i=1}^{m} (\bar{x}_i - x)^2, \tag{6.9}$$

and

$$s^2(\tilde{x}) = \frac{1}{m(m-1)} \sum_{i=1}^{m} (\bar{x}_i - \tilde{x})^2. \tag{6.10}$$

The estimator $\tilde{x}$, given by (6.8), is described as being based on *probability proportional to size (pps) estimation*; it is a serious competitor to the three described in the discussion of one-stage cluster sampling given in Section 5.3. Since the sampling has taken place with replacement, we must expect the procedure to be somewhat wasteful of effort: reflected in the value of $\mathrm{Var}(\tilde{x})$. But even so, circumstances can arise (in particular when the $\bar{X}_i$ and N_i are more or less uncorrelated) in which $\tilde{x}$ has similar efficiency to $\bar{x}_{C(a)}$, and both are more efficient than $\bar{x}_{C(b)}$ or $\bar{x}_{C(c)}$. $\tilde{x}_T$ has the added advantages of unbiasedness and ease of calculation, although the sampling procedure is more difficult and can be more costly (in view of the emphasis on the larger clusters). A more detailed comparison of these estimators is given by Cochran (1963, Chapter 9).

Sometimes this approach may need to be modified. If the cluster sizes N_i are not known precisely, then it may be necessary to use an alternative measure of cluster 'size', perhaps an estimate of N_i or some other quantity likely to be postively correlated with the N_i. For example, suppose we wish to sample primary schools throughout the country by taking a one-stage cluster sample of local education authorities. If knowledge of the numbers of schools for each authority happened not to be readily available, certain other factors might well be: such as expenditure on school education in the regions covered by the authorities, or the total populations in these regions. Either of these will be highly correlated with the number of schools, and will constitute a reasonable basis for probability sampling. Such sampling is now referred to as *sampling with probability proportional to estimated size* (or *ppes sampling*).

So we now assume that each cluster has such a measure, Z_i, associated with it, and use this as the basis for *ppes* sampling, which consists of choosing m clusters, with replacement, where at each stage of drawing the sample, cluster i has probability $p_i = Z_i/Z_T$ of being chosen ($i = 1, 2, \ldots, M; Z_T = \sum_{i=1}^{M} Z_i$).

X_T and $\bar{X}$ are now estimated by

$$\frac{Z_T}{m} \sum_{i=1}^{m} x_{iT}/Z_i \quad \text{and} \quad \frac{Z_T}{mN} \sum_{i=1}^{m} x_{iT}/Z_i,$$

respectively. These estimators are *unbiased*, and (by (6.2)) have variances

$$\frac{1}{m} \left\{ \sum_{i=1}^{M} \frac{Z_T X_{iT}^2}{Z_i} - X_T^2 \right\} \quad \text{and} \quad \frac{1}{m} \left\{ \sum_{i=1}^{M} \frac{Z_T X_{iT}^2}{N^2 Z_i} - \bar{X}^2 \right\},$$

respectively.

In the light of what we saw of the effect of sampling with probability proportional to the cluster total, the most advantageous measure of size in *ppes* sampling will clearly be that which is most nearly proportional to the cluster total.

6.4 Sampling without Replacement with Arbitrary Probabilities

The logical next step in complexity of sampling method is to consider the effects of sampling *without replacement*, and with arbitrary selection probabilities. We would expect that ruling out the possibility of multiple occurrence of population members in the sample must potentially lead to greater efficiency of estimation. But at the same time the derivation of the properties of estimators based on such sampling schemes can involve very complicated mathematical analyses, which are beyond the scope of this book. The large amount of fundamental research effort over recent years includes work on this topic, and on the incorporation (in the design and analysis of sample surveys) of any relevant circumstantial information beyond the range of direct sample data. This latter consideration leads to the application of Bayesian statistical principles in sampling theory.

The one form of *without-replacement sampling* which is easily handled is that where successive sample members are chosen at random from the residual population. This is just *simple random sampling*, which has served as the basic method of sampling underlying almost all the sampling schemes described in this book. Simple random sampling, in this sense, is encountered as overwhelmingly the most common principle in the day-to-day application of sampling methods for the study of finite populations.

Postscript

Throughout the earlier Chapters, the *Class Example* data (presented in Figure 1 of Section 1.6) has been used time and again to illustrate the properties of estimators for the different sampling methods that have been described. Depending on the structure of the population being investigated, on the existence of concomitant information on auxiliary variables or on prevailing cost information or administrative considerations, the different estimators were seen on theoretical grounds to have different relative advantages. These were conveniently illustrated by constructing approximate sampling distributions for large numbers of samples drawn from the *Class Example* population. We were able to see in practice how circumstances affected the properties of estimators derived from overall s.r. sampling, from the ratio method, or from stratified, or cluster, sampling.

As an *aide-memoire* to many of the results in this book, Figure 11 provides a summary of the *Class Example* results by presenting together all of the derived sampling distributions which have earlier been shown individually in their respective contexts. The reader can obtain a useful check on his understanding of the broad factors that support the use of different estimators of the population mean $\bar{X}$, by attempting to explain the reasons why the different sampling distributions shown in Figure 11 appear as they do in relation to one another.

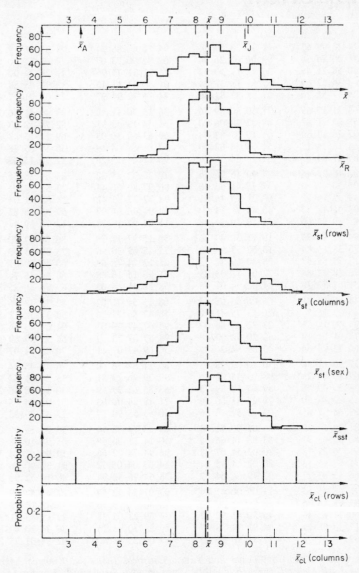

Figure 11 Summary of sampling distributions of estimators of $\bar{X}$ in the *Class Example*.

Appendix

53 74 23 99 67	61 32 28 69 84	94 62 67 86 24	98 33 41 19 95
63 38 06 86 54	99 00 65 26 94	02 82 90 23 07	79 62 67 80 60
35 30 58 21 46	06 72 17 10 94	25 21 31 75 96	49 28 24 00 49
63 43 36 82 69	65 51 18 37 88	61 38 44 12 45	32 92 85 88 65
98 25 37 55 26	01 91 82 81 46	74 71 12 94 97	24 02 71 37 07
02 63 21 17 69	71 50 80 89 56	38 15 70 11 48	43 40 45 86 98
64 55 22 21 82	48 22 28 06 00	61 54 13 43 91	82 78 12 23 29
85 07 26 13 89	01 10 07 82 04	59 63 69 36 03	69 11 15 83 80
58 54 16 24 15	51 54 44 82 00	62 61 65 04 69	38 18 65 18 97
34 85 27 84 87	61 48 64 56 26	90 18 48 13 26	37 70 15 42 57
03 92 18 27 46	57 99 16 96 56	30 33 72 85 22	84 64 38 56 98
62 95 30 27 59	37 75 41 66 48	86 97 80 61 45	23 53 04 01 63
08 45 93 15 22	60 21 75 46 91	98 77 27 85 42	28 88 61 08 84
07 08 55 18 40	45 44 75 13 90	24 94 96 61 02	57 55 66 83 15
01 85 89 95 66	51 10 19 34 88	15 84 97 19 75	12 76 39 43 78
72 84 71 14 35	19 11 58 49 26	50 11 17 17 76	86 31 57 20 18
88 78 28 16 84	13 52 53 94 53	75 45 69 30 96	73 89 65 70 31
45 17 75 65 57	28 40 19 72 12	25 12 74 75 67	60 40 60 81 19
96 76 28 12 54	22 01 11 94 25	71 96 16 16 88	68 64 36 74 45
43 31 67 72 30	24 02 94 08 63	38 32 36 66 02	69 36 38 25 39
50 44 66 44 21	66 06 58 05 62	68 15 54 35 02	42 35 48 96 32
22 66 22 15 86	26 63 75 41 99	58 42 36 72 24	58 37 52 18 51
96 24 40 14 51	23 22 30 88 57	95 67 47 29 83	94 69 40 06 07
31 73 91 61 19	60 20 72 93 48	98 57 07 23 69	65 95 39 69 58
78 60 73 99 84	43 89 94 36 45	56 69 47 07 41	90 22 91 07 12
84 37 90 61 56	70 10 23 98 05	85 11 34 76 60	76 48 45 34 60
36 67 10 08 23	98 93 35 08 86	99 29 76 29 81	33 34 91 58 93
07 28 59 07 48	89 64 58 89 75	83 85 62 27 89	30 14 78 56 27
10 15 83 87 60	79 24 31 66 56	21 48 24 06 93	91 98 94 05 49
55 19 68 97 65	03 73 52 16 56	00 53 55 90 27	33 42 29 38 87
53 81 29 13 39	35 01 20 71 34	62 33 74 82 14	53 73 19 09 03
51 86 32 68 92	33 98 74 66 99	40 14 71 94 58	45 94 19 38 81
35 91 70 29 13	80 03 54 07 27	96 94 78 32 66	50 95 52 74 33
37 71 67 95 13	20 02 44 95 94	64 85 04 05 72	01 32 90 76 14
93 66 13 83 27	92 79 64 64 72	28 54 96 53 84	48 14 52 98 94
02 96 08 45 65	13 05 00 41 84	93 07 54 72 59	21 45 57 09 77
49 83 43 48 35	82 88 33 69 96	72 36 04 19 76	47 45 15 18 60
84 60 71 62 46	40 80 81 30 37	34 39 23 05 38	25 15 35 71 30
18 17 30 88 71	44 91 14 88 47	89 23 30 63 15	56 34 20 47 89
79 69 10 61 78	71 32 76 95 62	87 00 22 58 40	92 54 01 75 25

This table is taken from Fisher and Yates: *Statistical Tables for Biological, Agricultural and Medical Research*, published by Longman Group Ltd., London (previously published by Oliver & Boyd, Edinburgh), and by permission of the authors and publishers.

Table 1   Random digits

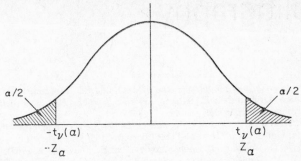

t-Distribution

Degrees of Freedom	α 0·10	0·05	0·02	0·01	0·002	0·001
ν						
1	6·314	12·71	31·82	63·66	318·3	636·6
2	2·920	4·303	6·965	9·925	22·33	31·60
3	2·353	3·182	4·541	5·841	10·22	12·94
4	2·132	2·776	3·747	4·604	7·173	8·610
5	2·015	2·571	3·365	4·032	5·893	6·859
6	1·943	2·447	3·143	3·707	5·208	5·959
7	1·895	2·365	2·998	3·499	4·785	5·405
8	1·860	2·306	2·896	3·355	4·501	5·041
9	1·833	2·262	2·281	3·250	4·297	4·781
10	1·812	2·228	2·764	3·169	4·144	4·587
12	1·782	2·179	2·681	3·055	3·930	4·318
14	1·761	2·145	2·624	2·977	3·787	4·140
16	1·746	2·120	2·583	2·921	3·686	4·015
18	1·734	2·101	2·552	2·878	3·611	3·922
20	1·725	2·086	2·528	2·845	3·552	3·850
25	1·708	2·060	2·485	2·787	3·450	3·725
30	1·697	2·042	2·457	2·750	3·385	3·646
40	1·684	2·021	2·423	2·704	3·307	3·551
60	1·671	2·000	2·390	2·660	3·232	3·460
80	1·664	1·990	2·374	2·639	3·195	3·415
∞	1·645	1·960	2·326	2·576	3·090	3·291

Normal Distribution

α	0·10	0·05	0·02	0·01	0·002	0·001
z_α	1·645	1·960	2·326	2·576	3·090	3·291

Table 2 Double-tailed percentage points, $t_\nu(\alpha)$ and z_α, for Student's t-distribution, and the Normal distribution.

Bibliography

Atkinson, J., *Handbook for Interviewers*. HMSO, 2nd Ed. (London, 1971).

Bartholomew, D. J. and E. E. Bassett, *Let's Look at the Figures*. Penguin Books (Harmondsworth, 1971).

Chung, J. H. and D. B. De Lury, *Confidence Limits for the Hypergeometric Distribution*. University of Toronto Press (Toronto, 1950).

Cochran, W. G., *Sampling Techniques*. Wiley, 2nd Ed. (New York, 1963).

Deming, W. E., *Sample Design in Business Research*. Wiley (New York, 1960).

Festinger, L. and D. Katz (Editors), *Research Methods in the Behavioural Sciences*. Holt, Rinehart and Winston (New York, 1953).

Fisher, R. A. and F. Yates, *Statistical Tables for Biological, Agricultural and Medical Research*. Longman Press, Reprint of 6th Ed. (Harlow, 1973).

Hansen, M. H., W. N. Hurwitz, and W. G. Madow, *Sample Survey Methods and Theory; Vol. 1 Methods and Applications, Vol. 2 Theory*. Wiley (New York, 1953).

Huff, D., *How to Lie with Statistics*. Penguin Books (Harmondsworth, 1973).

Hyman, H. H., *Interviewing in Social Research*. University of Chicago Press (Chicago, 1954).

Johnson, N. L. and H. Smith, *New Developments in Survey Sampling*. Wiley (New York, 1969).

Kish, L., *Survey Sampling*. Wiley (New York, 1965).

Moroney, M. J., *Facts from Figures*. Penguin Books (Harmondsworth, 1951).

Moser, C. A. and G. Kalton, *Survey Methods in Social Investigation*. Heinemann (London, 1971).

Nunnally, J. C., *Psychometric Theory*. McGraw–Hill (New York, 1967).

Owen, D. B., *Handbook of Statistical Tables*. Pergamon (London, 1962).

Parten, M. B., *Surveys, Polls and Samples*. Cooper Square (New York, 1966).

Payne, S. L., *The Art of Asking Questions*. Princton University Press (Princeton, 1951).

Raj, D., *Sampling Theory*. McGraw–Hill (New York, 1968).

Raj, D., *Design of Sample Surveys*. McGraw–Hill (New York, 1972).

Sampford, M. R., *An Introduction to Sampling Theory with Applications to Agriculture*. Oliver and Boyd (Edinburgh, 1962).

Savage, R. D. (Ed.), *Readings in Clinical Psychology*. Pergamon (London, 1966).

Slonim, M. J., *Sampling in a Nutshell*. Simon and Schuster (New York, 1960).

Social and Community Planning Research:
Technical Manual No. 1, Postal Survey Methods;
Technical Manual No. 2, Sample Design and Selection;
Technical Manual No. 3, Interviewers' Guidebook.
Social and Community Planning Research (London, 1972).

Stephan, F. F. and P. J. McCarthy, *Sampling Opinions*. Wiley (New York, 1958).

Stuart, A., *Basic Ideas of Scientific Sampling*. Griffin (London, 1962).

Yamane, T., *Elementary Sampling Theory*. Prentice–Hall (Eaglewood Cliffs, New Jersey, 1967).

Yates, F., *Sampling Methods for Censuses and Surveys*. Griffin, 3rd Ed. (London, 1960).

Young, P. V., *Scientific Social Surveys and Research*. Prentice–Hall, 4th Ed. (Eaglewood Cliffs, New Jersey, 1966).

Index

In this Author and Subject Index, authors' names are presented in CAPITAL letters. Subjects are referenced, where appropriate, under several headings. Where one heading provides greater detail, cross-references are given from other headings. Where subjects are discussed over several consecutive pages, only the *first* page number is given. Page numbers in **bold type** refer to major development of the subject.